Wen Liang is an associate professor at the School of Information Science and Engineering, Shenyang Ligong University, China. He holds a Ph.D. in Engineering from the Changchun University of Science and Technology, China. His research interests include complex networks and quantum walks.

Fangyan Dong is a professor at the Faculty of Mechanical Engineering and Mechanics, Ningbo University, China. She holds a Ph.D. in Engineering from the Tokyo Institute of Technology, Japan. Her research interests include computational intelligence, fuzzy systems, and quantum computing.

Exploring Complex Networks with Quantum Walks

Fei Yan, Wen Liang, and Fangyan Dong

CRC Press
Taylor & Francis Group
Boca Raton London New York

CRC Press is an imprint of the
Taylor & Francis Group, an **informa** business

SCIENCE PRESS

First edition published 2026
by CRC Press
2385 NW Executive Center Drive, Suite 320, Boca Raton FL 33431

and by CRC Press
4 Park Square, Milton Park, Abingdon, Oxon, OX14 4RN

CRC Press is an imprint of Taylor & Francis Group, LLC

ISBN: 978-1-041-15384-9 (hbk)
ISBN: 978-1-041-16323-7 (pbk)
ISBN: 978-1-003-68390-2 (ebk)

DOI: 10.1201/9781003683902

Typeset in Minion
by codeMantra

Contents

Figures

Tables

Acknowledgments

Figure 1.2 is reprinted from *Quantum Walks and Search Algorithms*, ISBN: 978-3-319-97813-0, Copyright (2018), with permission from Springer Nature.

Figure 1.3 is reprinted from *Physical Review A*, 79(5): 052335, Copyright (2009), with permission from American Physical Society.

Figures 3.9–3.11 are reprinted from *Neural Computing and Applications*, 34(16): 13455–13468, Copyright (2022), with permission from Springer Nature.

Figures 4.1–4.7 are reprinted from *Computer Communications*, 193: 378–387, Copyright (2022), with permission from Elsevier.

Figure 4.8 is reprinted from *Physical Review A*, 107(3): 032605, Copyright (2023), with permission from American Physical Society.

Table 3.2 is reprinted from *Lecture Notes in Computer Science*, 8621: 103–112, Copyright (2014), with permission from Springer Nature.

Tables 6.1 and 6.2 are reprinted from *Applied Intelligence*, 51(4): 2574–2588, Copyright (2020), with permission from Springer Nature.

Table 6.3 is reprinted from *Applied Intelligence*, 52(2): 1493–1507, Copyright (2021), with permission from Springer Nature.

Table 6.5 is reprinted from *IEEE Transactions on Neural Networks and Learning Systems*, 33(1): 292–303, Copyright (2022), with permission from IEEE.

This work is sponsored by the Natural Science Foundation of Liaoning Province, China (2024-BS-113), the Projects Proposed for Independent Research Topics by the Education Department of Liaoning Province, China (LJ212410144003), and the Natural Science Foundation of Jilin Province, China (20240101332JC).

The authors acknowledge Hesheng Huang, Shuyu Liu, and Haoyu Wang for their assistance with data collection, literature organization, and the translation of specific portions.

I

Theory

CHAPTER 1

Quantum Computing and Quantum Walks

D URING THE LATE SPRING and Autumn Period and early Warring
States Period, the ancient Chinese philosopher Mozi defined "force"
in his work *Mozi's Canon* as "force is what causes matter to move," revealing that force is the reason for the movement of objects. Humans have per-
sistently explored mechanics, and it was only in the 19th century that the
perspective of mechanical research in the macro world shifted to the micro
world. In the early 20th century, quantum mechanics, a branch of physics that studies the motion of microscopic particles, was established. The
birth of quantum mechanics led humans to develop new thinking about
the microscopic world and even the universe. In 1981, based on the theory
of quantum mechanics, the Argonne National Laboratory in the United
States proposed quantum computing, giving quantum mechanics a new
life with characteristics of the information age. Moore's Law states that the
number of transistors that can be accommodated on an integrated circuit
will double every 18 months (or every two years). However, physical components cannot be shrunk indefinitely, and Moore's Law will eventually
reach its end. There is an inevitability in dedicating efforts to the research
and development of quantum computing and quantum computers to align
with future technological trends. Currently, quantum computing exhibits a
"dual-track parallel" development model. On one hand, quantum computers are undergoing diverse research and development efforts, with superconducting quantum, ion traps, diamond color centers, nuclear magnetic

DOI: 10.1201/9781003683902-2

resonance, D-wave annealing machines, and linear optics emerging as mainstream materials and technologies for quantum computer research and development. On the other hand, quantum algorithms designed to run on quantum computers are flourishing and competing with each other. In 1994, Shor proposed Shor's algorithm, a quantum algorithm for prime factorization [1]. Because this algorithm has only polynomial time complexity, it poses a serious challenge to the security of modern cryptography, which relies on the difficulty of factoring large prime numbers. At the same time, it provides a strong research impetus for the design of quantum computing and quantum algorithms. In 1996, Grover proposed a quantum version of the database search algorithm, which achieves quadratic speedup compared to classical algorithms [2]. He also formalized the important technique of amplitude amplification in quantum algorithm design, which has significantly advanced research on quantum walk algorithms for marked point search problems. In 2009, the HHL (Harrow, Hassidim, and Lloyd) quantum algorithm became known for its ability to save exponential time in solving linear equations [3], greatly promoting the development of quantum machine learning [4–6].

In recent years, quantum computing has been deployed in various fields, sparking a series of groundbreaking explorations into its future applications. For instance, Fujitsu Group has adopted a quantum-inspired annealing algorithm to assist Nippon Yusen Kabushiki Kaisha (NYK Line) in optimizing complex vessel stowage planning problems, aiming to improve scheduling efficiency. This project was first put into use in September 2021, and NYK Line predicts that the adoption of the quantum-inspired annealing algorithm will save approximately 4,000 hours of vessel stowage time annually [7]. Beijing ByteDance Technology Co., Ltd. disclosed theoretical achievements in simulating the properties of small chemical molecules using quantum computing in 2022, which are expected to play a positive role in promoting the industrial chemistry field [8]. Hefei Origin Quantum Computing Technology Co., Ltd., in collaboration with China Construction Bank, has launched a quantum option pricing application and a quantum VaR (Value at Risk) calculation algorithm, which offers superior speed and accuracy compared to similar international algorithms [9]. Quantum finance applications, represented by quantum VaR calculations, are expected to attract significant capital into quantum computing and prioritize rapid deployment in fields such as biology, medicine, and education, injecting new vitality into the research, development, and application of quantum computing.

Moreover, internationally renowned companies such as Google and IBM are committed to developing quantum computers and have hyped the concept of "quantum supremacy" (more accurately, quantum advantage) through social media, intensifying the urgency of quantum computing research and development. In China, companies represented by Baidu, Alibaba, Tencent, and Huawei are racing to deploy quantum computing to tap into its vast application potential. Additionally, teams represented by Academician Pan Jianwei have made frequent achievements in the fields of quantum walks [10] and ultra-long-distance quantum key distribution [11] in recent years. Moreover, the 14th Five-Year Plan for National Economic and Social Development of the People's Republic of China explicitly identifies quantum information as a strategic and forward-looking frontier area, significantly raising public awareness of quantum computing. In the future, quantum computing will play a significant role in various fields such as military, chemistry, biology, information security, and finance [5,12,13].

This chapter covers the basic concepts and commonly used operations involved in quantum computing, laying the foundation for subsequent research on the application of quantum walks in complex networks.

1.1 BASIC CONCEPTS OF QUANTUM COMPUTING

This section covers Dirac notation and qubits, common operations and operators, basic concepts of quantum circuits, and the fundamental assumptions of quantum mechanics.

1.1.1 Dirac Notation and Qubits

Dirac notation (also known as bra-ket notation) is a specific expression for vectors in quantum mechanics. For example, a vector u corresponds to the Dirac notation as

$$|u\rangle = (a_1, \cdots, a_n)^{\mathrm{T}}, \tag{1.1}$$

$$\langle u| = (a_1^*, \cdots, a_n^*), \tag{1.2}$$

where T represents the transpose symbol; $\langle u|$ signifies the conjugate transpose of $|u\rangle$; a_1^* indicates the conjugate value of the element a_1, which only takes effect when it contains imaginary numbers. In practical computations, the calculation of $|u\rangle$ may be simpler, similar to shallow embedding in network representation learning or the one-hot encoding method in graph neural networks. When $|u\rangle$ has only one element equal to 1, it is called an orthogonal basis or computational basis, collectively referred to as the standard basis in this book. Assuming that in a complex network with n nodes, any node is expressed in the form of a standard basis, then

$$|1\rangle = \begin{pmatrix} 1 \\ 0 \\ \vdots \\ 0 \end{pmatrix}, \ |2\rangle = \begin{pmatrix} 0 \\ 1 \\ \vdots \\ 0 \end{pmatrix}, \ \cdots, \ |n\rangle = \begin{pmatrix} 0 \\ 0 \\ \vdots \\ 1 \end{pmatrix}. \tag{1.3}$$

In this chapter, the standard basis corresponding to network nodes and the coin state in quantum walks are both expressed in the form of Eq. (1.3).

The qubit, the smallest unit of information in quantum computing, is represented as a linear combination of unit vectors, also known as a qubit. Any qubit is represented using Dirac notation. Given $|0\rangle = (1,0)^{\mathrm{T}}$ and $|1\rangle = (0,1)^{\mathrm{T}}$, a quantum state composed of $|0\rangle$ and $|1\rangle$ qubits can be defined as

$$|\psi\rangle = \alpha|0\rangle + \beta|1\rangle = \alpha \begin{pmatrix} 1 \\ 0 \end{pmatrix} + \beta \begin{pmatrix} 0 \\ 1 \end{pmatrix}, \tag{1.4}$$

where α and β indicate the probability amplitudes, which are generally expressed in complex numbers. In classical bits, 0 and 1 can represent information with oppositional relationships, such as true and false. In qubits, however, $|0\rangle$ and $|1\rangle$ are "contradictory and unified entities" because a qubit can exist in a superposition state containing both 0 and 1 simultaneously. This coding method in qubits offers a richer set of information, which is the primary reason why qubits possess the potential for parallelism. Taking the quantum (inspired) swarm intelligence algorithm as an example [14], the dual-chain coding method based on the $|0\rangle$ and $|1\rangle$ forms for the initial population can avoid local optima during the search process. In the quantum walks discussed in this book, $|0\rangle$ and $|1\rangle$, respectively, represent the coin states of heads up or down after flipping a coin. In other quantum algorithms, $|0\rangle$ and $|1\rangle$ can distinguish target and non-target data. Moreover, in quantum circuits, $|0\rangle$ and $|1\rangle$ can act as control bits to influence the computation of operators, outputting the desired results. In

Eq. (1.4), the linear combination of the vectors $|0\rangle$ and $|1\rangle$ is referred to as a superposition state. When $\alpha = \beta = 1/\sqrt{2}$, $|\psi\rangle$ is said to be in an equal superposition state. In an isolated quantum system, the following condition is always satisfied:

$$|\alpha|^2 + |\beta|^2 = 1, \qquad (1.5)$$

where $|\cdot|^2$ represents the modulus squared operation for complex numbers.

The more generalized form of Eq. (1.4) places the quantum state within the Bloch sphere, introducing the angle θ and the phase parameter φ, such that $|\psi\rangle = \cos(\theta/2)|0\rangle + e^{i\varphi} \cdot \sin(\theta/2)|1\rangle$, where i is the imaginary unit. Since the core content of this book involves little discussion of this part of the knowledge, we will not elaborate further. Interested readers may refer to the detailed introductions found in references [13,15].

1.1.2 Common Operations and Operators

1.1.2.1 Inner Product and Outer Product

For vectors and quantum states represented using Dirac notation, common operations in quantum computation include the inner product and outer product. Taking $|u\rangle$ from Eq. (1.1) and $\langle u|$ from Eq. (1.2) as examples, the result of the inner product corresponds to a value, for example,

$$\langle u|u \rangle = \sum_{i=1}^{n} a_i^* a_i. \qquad (1.6)$$

The outer product of $|u\rangle$ and $\langle u|$ is represented by

$$|u\rangle\langle u| = \begin{pmatrix} a_1 \\ \vdots \\ a_n \end{pmatrix} \cdot \left(a_1^*, \cdots, a_n^* \right) = \begin{pmatrix} a_1 a_1^* & \cdots & a_1 a_n^* \\ & \ddots & \\ a_n a_1^* & \cdots & a_n a_n^* \end{pmatrix}. \qquad (1.7)$$

In this book, the inner product operation is primarily applied during the quantum measurement stage, where the measurement results from the quantum walk algorithm are used to score the network nodes and links. The outer product operation is primarily used to construct the evolution operator, which drives discrete-time quantum walks or specifies the walk paths for particles. Additional details on quantum measurement and evolution are provided in Section 1.1.4.

1.1.2.2 Tensor Product and Direct Sum

When the objects of operation are matrices or vectors, the more widely used operations in quantum computation are the tensor product and the direct sum. The tensor product is represented by the symbol $\otimes$. Suppose there are matrices $A_{2\times1}$ and $B_{2\times3}$ with dimensions 2×1 and 2×3, respectively, their tensor product can be expressed as

$$
A_{2\times1} \otimes B_{2\times3} = \begin{pmatrix} a_1 \\ a_2 \end{pmatrix} \otimes \begin{pmatrix} b_{11} & b_{12} & b_{13} \\ b_{21} & b_{22} & b_{23} \end{pmatrix}
$$

$$
= \begin{pmatrix} a_1 b_{11} & a_1 b_{12} & a_1 b_{13} \\ a_1 b_{21} & a_1 b_{22} & a_1 b_{23} \\ a_2 b_{11} & a_2 b_{12} & a_2 b_{13} \\ a_2 b_{21} & a_2 b_{22} & a_2 b_{23} \end{pmatrix}.
$$

(1.8)

Similarly, the tensor product of matrix $A_{m\times n}$ and matrix $B_{p\times q}$, where the dimensions are $m \times n$ and $p \times q$, respectively, results in a matrix $C_{mp\times qn}$. It is worth noting that in quantum computing, the tensor product symbol is often omitted, such as

$$
|u\rangle \otimes |v\rangle = |u\rangle|v\rangle = |u,v\rangle.
$$

(1.9)

Regarding the direct sum, the symbol "$\oplus$" is used throughout the book. Taking the matrices $A_{2\times1}$ and $B_{2\times3}$ as examples again, their direct sum can be expressed as

$$
A_{2\times1} \oplus B_{2\times3} = \begin{pmatrix} a_1 \\ a_2 \end{pmatrix} \oplus \begin{pmatrix} b_{11} & b_{12} & b_{13} \\ b_{21} & b_{22} & b_{23} \end{pmatrix}
$$

$$
= \begin{pmatrix} a_1 & 0 & 0 & 0 \\ a_2 & 0 & 0 & 0 \\ 0 & b_{11} & b_{12} & b_{13} \\ 0 & b_{21} & b_{22} & b_{23} \end{pmatrix} = \begin{pmatrix} A_{2\times1} & 0 \\ 0 & B_{2\times3} \end{pmatrix}.
$$

(1.10)

Obviously, the direct sum of matrices $A_{m \times n}$ and $B_{p \times q}$, which have dimensions $m \times n$ and $p \times q$, respectively, results in a matrix $C_{(m+p) \times (n+q)}$. In this book, tensor product and direct sum operations are primarily used to construct the Hilbert space of closed quantum systems. Additional details on Hilbert space are provided in Section 1.1.4.

1.1.2.3 Three Common Operators

Below, we introduce three common operators, that is, the Hadamard operator, the Fourier operator, and the Grover operator. The expression of the Hadamard operator is relatively simple, which can be viewed as the summation of the outer product results of the qubits $|0\rangle$ and $|1\rangle$:

$$H = \frac{1}{\sqrt{2}} \left(|0\rangle\langle 0| + |0\rangle\langle 1| + |1\rangle\langle 0| - |1\rangle\langle 1| \right)$$

$$= \frac{1}{\sqrt{2}} \begin{pmatrix} 1 & 1 \\ 1 & -1 \end{pmatrix}. \tag{1.11}$$

The Fourier operator can be regarded as a matrix controlled by a phase parameter ω. Let the phase parameter $\omega = \exp(2\pi i/N)$, where N is the number of rows and columns of the operator, and i is the imaginary unit. The calculation method for the element in the pth row and qth column of the Fourier operator F is as follows:

$$F_{pq} = \frac{\omega^{pq}}{\sqrt{N}} = \frac{\left[\exp(2\pi i/N)\right]^{pq}}{\sqrt{N}}, \tag{1.12}$$

where p and q start counting from 0, with both belonging to the set $\{0,\cdots,N-1\}$. Due to the inclusion of the imaginary unit i, Eq. (1.12) will introduce Euler's formula during the calculation, transforming the original expression into trigonometric functions. The Euler's formula is defined as

$$\exp(i\theta) = \cos(\theta) + i \cdot \sin(\theta). \tag{1.13}$$

Assuming the Fourier operator has a matrix size of 4, and combining Eqs. (1.12) and (1.13), the Fourier operator F is formulated as

$$F = \frac{1}{2} \begin{pmatrix} 1 & 1 & 1 & 1 \\ 1 & i & -1 & -i \\ 1 & -1 & 1 & -1 \\ 1 & -i & -1 & i \end{pmatrix}. \tag{1.14}$$

The Grover operator is defined based on the reflection operation of Grover search algorithm [2]. Additional details on Grover search algorithm and its reflection operator are provided in Section 1.2.1. A simple two-level Grover operator G can be defined as

$$G = \frac{1}{2} \begin{pmatrix} -1 & 1 & 1 & 1 \\ 1 & -1 & 1 & 1 \\ 1 & 1 & -1 & 1 \\ 1 & 1 & 1 & -1 \end{pmatrix}. \tag{1.15}$$

According to the expression in Eq. (1.15), the Grover operator appears as a matrix with negative values on the main diagonal. In this book, when the diagonal elements of the Grover operator incorporate attribute information of the objects under study, they can be used to design independent coin operators in quantum walk algorithms.

It is worth noting that the introduction of the phase parameters ϕ_1, ϕ_2, and the angle parameter θ represents a more general construction of unitary operators, such as the SU(2) operator [16]:

$$SU(2) = \begin{pmatrix} \cos(\theta) & e^{i\phi_1}\sin(\theta) \\ e^{i\phi_2}\sin(\theta) & -e^{i(\phi_1+\phi_2)}\cos(\theta) \end{pmatrix}. \tag{1.16}$$

The primary uses of the SU(2) operator include: (1) Discussion of the relationship between phase parameters and quantum measurement results. For instance, examining how the phase parameters within the SU(2) operator influence the longest walk distance in quantum walks based on information entropy [17]. (2) Construction of new unitary operators. As an example, when the angle parameter $\theta = \pi/4$, and the phase parameters $\phi_1 = \phi_2 = 0$ in the SU(2) operator, the operator becomes the Hadamard operator as described in Eq. (1.11). When the parameters θ, ϕ_1, and ϕ_2 are arbitrary, a unitary matrix can be obtained from Eq. (1.16). The concept of unitarity is referenced in Section 1.1.4. Additionally, quantum computing involves various logic gates such as NOT gates and SWAP gates, etc.

However, as this topic is not extensively covered in this book, it will not be discussed in detail here. Interested readers may access literature [15,18,19] for further information.

1.1.3 Basic Concepts of Quantum Circuits

Quantum circuits visually express the operations of quantum operators (logic gates) and show the complexity of quantum algorithms on circuit diagrams. Figure 1.1a depicts a simple quantum circuit diagram, where each rail (wire) represents a qubit, and each box with a letter represents an operator. The time flow of the quantum circuit moves from left to right, with the operations of the operators on the left executed before those on the right in chronological order. In Figure 1.1a, a single rail represents and transmits a qubit, while a double rail represents classical bits 0 and 1. Furthermore, in Figure 1.1, H, X, Y, Z, A, B, C, D, E, F, and U denote different operators. Operators on different rails perform tensor product operations, for example, the circuit in Figure 1.1b is represented as $A \otimes B \otimes C$. In contrast, operators on the same rail perform dot product operations according to the flow of time, where the dot product refers to the element-wise multiplication of two matrices with the same dimensions. For instance, the operators in the circuit of Figure 1.1c perform a dot product operation, expressed as $F \cdot E \cdot D$. It is important to note that in quantum computing, the symbols for tensor products and matrix multiplications

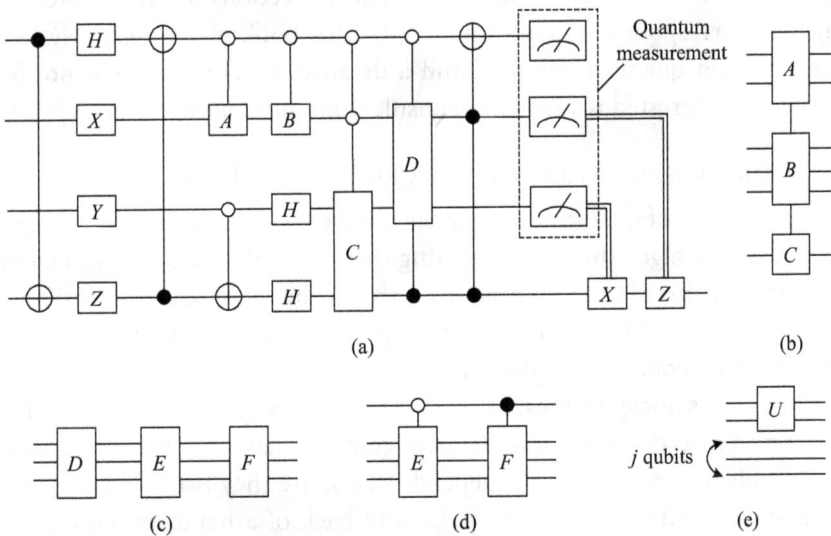

FIGURE 1.1   Examples of quantum circuits.

are often omitted. To distinctly differentiate between the two operations, the order of writing is required to be opposite to the order of matrix multiplication [20]. Matrix multiplication operations, including dot products, satisfy the commutative property, meaning that reversing the order of operations does not affect the calculation results.

In a quantum circuit diagram, a hollow circle represents a control bit of $|0\rangle$, while a solid black circle represents a control bit of $|1\rangle$. When a control bit on a certain rail is connected to an operator on an adjacent rail, the operation result of that operator is influenced by the control bit being either $|0\rangle$ or $|1\rangle$. Controlled operators (logic gates) perform direct sum operations. Taking the quantum circuit diagram in Figure 1.1d as an example, it illustrates $E \oplus F$. The operation of the quantum circuit is parallel, which can be reflected in the expression of the input qubits. According to Eq. (1.4), one qubit contains the information of two classical bits. In the quantum circuit diagram in Figure 1.1e, after j qubits perform tensor product operations, the number of classical bits contained is 2^j. When there are no operators present on the rail of the qubits, an identity matrix I is assumed to occupy that position, thus the quantum circuit in Figure 1.1e is expressed as $U \otimes I_{2^j}$, where I_{2^j} denotes the identity matrix of size 2^j.

General quantum algorithms refer to algorithms that can run on a quantum computer, while quantum circuits are the model basis for quantum algorithms to operate on quantum devices. Even if the three quantum algorithms introduced in the subsequent sections of this chapter all include corresponding quantum circuits, this book places relatively low emphasis on quantum circuits, and a detailed introduction will not be provided. Interested readers may consult *Quantum Image Processing* [12].

1.1.4 Fundamental Assumptions of Quantum Mechanics

The postulates of quantum mechanics are the core foundation for designing quantum algorithms and enabling their potential parallelism. Before introducing these basic assumptions, let's start by interpreting the cultural treasures of the Chinese nation to help readers quickly understand the fundamental concepts of quantum mechanics.

As early as ancient times, Fu Xi had formed a vague concept of binary through Yin and Yang. These elements contain each other to create a unified whole; neither are they independent, nor are they two separate parts of a single entity. Just like the palm and back of a hand, which can be described as Yin and Yang, respectively, but cannot be separated as a whole. Today, thousands of years later, the state vectors represented by qubits in quantum computing quantitatively express this intertwined concept of Yin and Yang. As presented in Eq. (1.4), $|0\rangle$ and $|1\rangle$ indicate Yin and Yang, while the probability amplitudes of $|0\rangle$ and $|1\rangle$ correspond to the dynamic changes of Yin–Yang harmony within an entity or system. This is also a major reason why Chinese people generally believe that quantum mechanics is closely related to traditional Chinese culture.

During the Western Han Dynasty, ancient Chinese people summarized the "24 solar terms" to guide farming activities based on their exploration of astronomy and calendar systems. The essence of the "24 solar terms" lies in the movement patterns of stars along the ecliptic longitude coordinate system over time. Similarly, Fu Xi's Eight Trigrams and Yao Ci in the *Book of Changes* symbolize the ancient Chinese people's exploration of evolution patterns under the astronomical calendar system. From the macro universe to the micro world, the same principle applies: uncovering the patterns of evolution over time within seemingly chaotic systems. In the process of exploring the vast and boundless universe, Wang Guowei's statement reveals the limitations and narrowness of human observation: "Viewing the world through oneself, everything becomes tinted with the hues of one's own thoughts and feelings." Each person's perception of the three thousand great worlds is only one facet of the complete picture. Just as in the micro world, humans can only observe the movement results of particles at a certain moment, unable to see the full picture at once. This leads to the important concept in quantum mechanics that was once questioned by Einstein, that is, collapse.

If the world is a high-dimensional information space, then what an individual observes is a projection of this high-dimensional information. As Mencius said, "The essence of all things resides within oneself." In real life, projection can be visualized as a person's shadow under the sun. Although the projection result is a reduced-dimension depiction of real information over time, many times we can still guess the posture, handheld items, body features, and even the shadow owner based on the shadow's outline characteristics. Therefore, measurements of individuals at a certain moment in quantum mechanics are not meaningless. Even if the measured result is only one facet of the complete picture of the micro world, it is still significant for the current moment. This is also the key idea of quantum algorithms to find target solutions from oscillating measurement results.

The above content covers several important concepts of quantum mechanics: evolution, projection, measurement, and collapse. Based on these popularized explanations, let's introduce the four basic assumptions of quantum mechanics: state space, unitary evolution, composite systems, and measurement processes [21]. In the context of complex network problems set by this book, the following explanations are provided for these four basic assumptions: (1) Treating complex networks as closed quantum systems, the state space is defined by considering the complex network under study as a quantum state $|\psi\rangle$ within a Hilbert space. The movement of particles in the Hilbert space is recorded through different states $|\psi(t)\rangle$ at various times t. When quantum walks occur on a specific network, particles can only jump between nodes based on their link relationships. However, this does not imply that the dimensionality of the state space is equal to the number of network nodes. Considering a one-dimensional line quantum walk as an example, since the particle has one direction to move left and another to move right, the spatial dimension of this line is defined as twice the number of position points on the line. (2) Unitary evolution, also known as unitary transformation, enumerates all possible movements of particles within a closed space over time. Unitary evolution shares similarities with the "24 solar terms" mentioned earlier in describing motion patterns. In quantum walks, unitary evolution emphasizes two constraints: first, the sum of probabilities for particles to stop at all nodes after measurement equals 1; second, during the evolution process (before measurement), the particle's motion situation before the current moment can be restored through the product of the inverse evolution operator and the state vector, for example, the calculation process of $|\psi(0)\rangle=|\psi(t=1)\rangle U^{-1}$ and $|\psi(t=1)\rangle=U|\psi(0)\rangle$. Unitary evolution relies on unitary matrices, which satisfy the following properties:

$$UU^{\dagger} = I, \tag{1.17}$$

or

$$U^{\dagger}=U^{-1}, \tag{1.18}$$

where I denotes the identity matrix, and $U^{\dagger}$ indicates the conjugate transpose matrix of U. Simply put, the state space represents the entirety of the closed quantum system currently under study, while unitary evolution describes the laws governing the motion of particles within that space. (3) In this book, a composite system specifically refers to the state space corresponding to complex networks, which is constructed from the tensor products, sums, or direct sums of the components of each node in the network. (4) The measurement process is a destructive observational action on the components of the state vector that are in a superposition state. In this book, measurement predominantly refers to calculating the staying probabilities of particles at different nodes within the complex network, which is then used as a scoring method for nodes and links in specific problems. Measurement carries the connotation of "Viewing the world through oneself, everything becomes tinted with the hues of one's own thoughts and feelings." It reflects the projection of the components of the state vector onto the evolution operator, usually defined as

$$P = \left| \langle \psi | U | \psi \rangle \right|^2 . \tag{1.19}$$

Using the complex network problem environment as an example, Eq. (1.19) represents the probability distribution of a particle staying on all nodes of the network after the evolution of the operator U over time.

Quantum algorithms are programs (circuits) designed to be run on quantum devices based on the fundamental assumptions of quantum mechanics mentioned above. The main idea is to encode the problem into quantum states, adjust the probability amplitudes of various components during evolution, and induce the splitting of the entire superposition state. Finally, measurement produces the output as target solutions and non-target solutions.

1.2 INTRODUCTION TO QUANTUM ALGORITHMS

Research in quantum algorithms has seen significant advances, giving rise to a variety of performance-optimized algorithms such as Grover search algorithm [2], variational quantum algorithms [22], HHL quantum algorithms [3], quantum support vector machines [23], quantum singular value decomposition [24], quantum Fourier transform [1], amplitude estimation [25], amplitude amplification [2], adiabatic quantum computing [26], and quantum walks [21]. This section introduces three representative algorithms, including Grover search algorithm, quantum walks, and HHL quantum algorithms.

1.2.1 Grover Search Algorithm

The Grover search algorithm [2] enables fast searching of unstructured data. For instance, when searching for someone's phone number in a telephone directory with N disordered entries, Grover search algorithm can achieve search acceleration with a complexity of $O(\sqrt{N})$ compared to classical search algorithms. When the amount of data in the database to be searched is extremely large, the resources saved by the square-root acceleration are considerable. The core idea of the Grover search algorithm is as follows: suppose the database contains N data points, where the black box (oracle) classifies all data points into marked points and unmarked points. As the algorithm evolves, the sign of the amplitude corresponding to the marked point is flipped, resulting in an amplification of their amplitudes while reducing those of the unmarked points. Repeating the above process $\pi\sqrt{N}/4$ (rounded down) times will cause a measurement probability greater than or equal to $1-1/N$ for the marked point, while the measurement probability of other nodes tends to 0, thereby achieving fast searching.

The mathematical expression process of the Grover search algorithm is as follows [21]: assume there are $N=2^n$ data points, where n is the number of qubits associated with these N data points. The marked function $f(x)$ equals 1 only at the marked point $x = x_0$, otherwise, it equals 0. Therefore, $f(x)$ is defined as

$$f(x)=\begin{cases} 1, & \text{if } x = x_0 \\ 0, & \text{otherwise} \end{cases}. \tag{1.20}$$

In quantum computing, the evolution is the core step for the algorithm to achieve specific functions. To realize the marking function of the Grover search algorithm on data points under a superposition state, it is necessary to define an evolution operator U. This operator consists of unitary operators R_f and R_D, where R_f is the quantum version of the marked function $f(x)$. It indicates whether the currently evaluated data point x is the marked point x_0 through $|0\rangle$ and $|1\rangle$. Hence, R_f is defined as

$$R_f\,|x\rangle|0\rangle=\begin{cases} |x_0\rangle|1\rangle, & \text{if } x = x_0 \\ |x\rangle|0\rangle, & \text{otherwise} \end{cases}. \tag{1.21}$$

Equation (1.21) represents the oracle used in the Grover search algorithm. Let all data points be expressed in the form of a standard basis, and denoted as $|j\rangle$. This allows us to construct the superposition state of all data points at the initial moment as follows:

$$|D\rangle = \frac{1}{\sqrt{N}} \sum_{j=0}^{N-1} |j\rangle. \tag{1.22}$$

From this, a unitary operator R_D can be defined, with all diagonal elements being negative, also known as the reflection operator:

$$R_D = \left(2|D\rangle\langle D| - I_N\right) \otimes I_2, \tag{1.23}$$

where I_N and I_2 indicate the identity matrices with dimensions N and 2, respectively. In the Grover search algorithm, each search iteration involves the execution of the evolution operator U, which is described as

$$U = R_D R_f. \tag{1.24}$$

Furthermore, the probability amplitude of the target data can be quickly adjusted through sign flipping. As such, the initial state is set as $|-\rangle = (|0\rangle - |1\rangle)/\sqrt{2}$, and the quantum state at the initial moment is expressed as

$$|\psi_0\rangle = |D\rangle|-\rangle. \tag{1.25}$$

Suppose the black box R_f has detected the presence of a marked point $|x_0\rangle$. At this point, the sign-flipping effect of the state $|-\rangle$ becomes clear, that is,

$$R_f|x_0\rangle|-\rangle = \frac{R_f|x_0\rangle|0\rangle - R_f|x_0\rangle|1\rangle}{\sqrt{2}}$$

$$= \frac{|x_0\rangle|1\rangle - |x_0\rangle|0\rangle}{\sqrt{2}} \tag{1.26}$$

$$= -|x_0\rangle|-\rangle.$$

Through the transformation of Eq. (1.26), it is found that the initial state $|D\rangle$ of all data points must be supplemented with a state $|-\rangle$ to achieve amplitude amplification under the action of R_f, which is the sign flipping of the probability amplitude corresponding to the marked point.

For the sake of understanding, this section provides a simplified example to summarize the complex description above. Assume the database contains four elements, and the initial probability amplitudes of each element are recorded in a matrix y. The data point relevant to the third element is the marked target search data, and its probability amplitude is marked in matrix y with a box, denoted as

$$y = \frac{1}{2}\begin{pmatrix} 1 \\ 1 \\ \boxed{1} \\ 1 \end{pmatrix}. \tag{1.27}$$

Following the above, Eq. (1.21) is employed to flip the sign of the probability amplitude corresponding to the marked point, resulting in:

$$y \cdot R_f = \begin{pmatrix} 1 & 0 & 0 & 0 \\ 0 & 1 & 0 & 0 \\ 0 & 0 & -1 & 0 \\ 0 & 0 & 0 & 1 \end{pmatrix} \times \frac{1}{2}\begin{pmatrix} 1 \\ 1 \\ \boxed{1} \\ 1 \end{pmatrix} = \frac{1}{2}\begin{pmatrix} 1 \\ 1 \\ \boxed{-1} \\ 1 \end{pmatrix}. \tag{1.28}$$

Equation (1.28) represents to the implementation of the sign-flipping function in Eq. (1.26). Finally, a Grover operator in the form of Eq. (1.15) is utilized to achieve amplitude amplification, which can be described as the product of Eq. (1.28) and the Grover operator:

$$(y \cdot R_f) \cdot G = \frac{1}{2}\begin{pmatrix} -1 & 1 & 1 & 1 \\ 1 & -1 & 1 & 1 \\ 1 & 1 & -1 & 1 \\ 1 & 1 & 1 & -1 \end{pmatrix} \times \frac{1}{2}\begin{pmatrix} 1 \\ 1 \\ \boxed{-1} \\ 1 \end{pmatrix} = \begin{pmatrix} 0 \\ 0 \\ \boxed{1} \\ 0 \end{pmatrix}. \tag{1.29}$$

At this point, it can be seen that only the probability amplitude corresponding to the marked point is greater than 0, while the amplitudes of all other non-marked points are equal to 0. This example intuitively demonstrates the core idea and calculation steps of the Grover search algorithm.

In addition, based on the above definitions, a quantum circuit diagram for the Grover search algorithm can also be designed to execute the Grover search algorithm on a quantum computer. In the quantum circuit for the Grover search algorithm, $|D\rangle$ represents the superposition state of n data points, where $n = \log_2 N$. According to the superposition state in Eq. (1.22), $|D\rangle$ is expressed in the quantum circuit diagram as

$$|D\rangle = H^{\otimes n}|0\rangle. \tag{1.30}$$

Furthermore, using an operator $\left(I - 2|0\rangle\langle 0|\right)$ to act as the marking function R_f in the quantum circuit for the Grover search algorithm, combined with Eqs. (1.21), (1.23), (1.24), and (1.25), R_D is defined as follows:

$$R_D = -\left(H^{\otimes n}\left(I - 2|0\rangle\langle 0|\right)H^{\otimes n}\right) \otimes I_2. \tag{1.31}$$

As a result, the quantum circuit diagram for the Grover search algorithm can refer to Figure 1.2, where the operator R_f playing the role of the black box and the reflection operator of the Grover search algorithm are marked with dashed boxes. The quantum circuit module inside the boxes runs $\pi\sqrt{N}/4$ times (this result is rounded down to the nearest integer).

Grover search algorithm offers valuable techniques for the design of quantum algorithms, that is, amplitude amplification. The mechanism of amplitude amplification has become a crucial step in many quantum algorithms, such as providing a quantum solution for the optimal arm selection in multi-armed bandit problems in reinforcement learning [27]. Additionally, the Grover search algorithm has led to the derivation of operators similar to Eq. (1.23), which provide new driving forces (coin operators) for the movement of particles on graphs in quantum walk algorithms. Particularly, the process of "embedding marked information

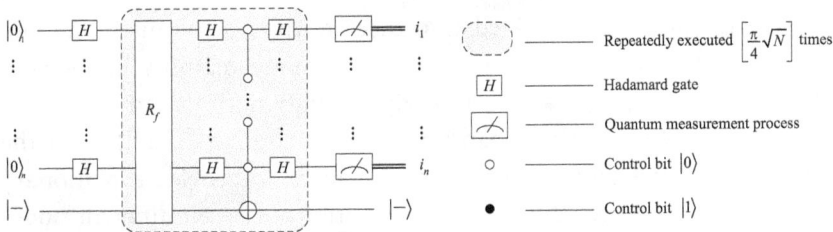

FIGURE 1.2   Quantum circuit diagram for the Grover search algorithm [21].

in the operator" and "amplifying the probability amplitude of the target solution through evolution" in the Grover search algorithm has already formed a mature quantum algorithm design philosophy [3], significantly advancing the research and innovation of quantum algorithms for spatial search problems [21,28].

1.2.2 Quantum Walks

Strictly speaking, without any limitations or requirements, quantum walks are considered computational models in quantum computing [29], because they can be used to design and implement other quantum algorithms. As an example, there is a quantum algorithm based on quantum walks designed to verify whether the product of two matrices equals a third matrix [30]. When quantum walks are employed to implement specific functions in practical applications, they can be identified as quantum walk algorithms. Revisiting the Grover search algorithm from Section 1.2.1, assuming that associations exist among all data points, the Grover search algorithm can be viewed as a search for marked nodes on a complete graph [31,32]. In contrast, quantum walks can be considered as a Grover search algorithm for structured data [33].

Quantum walks are extensions of classical random walks [34,35], but their characteristics are vastly different compared to classical random walks. The main distinction lies in the fact that the evolution process of quantum walks is not random; the only process where randomness may occur is during the measurement stage of the quantum walk, where the particle may randomly collapse onto a certain node in the graph. In the context of complex networks studied in this book, the node where the particle resides is directly specified during the measurement process, so quantum walks do not exhibit randomness. The main entity in quantum walks is the same as in classical random walks, both being walkers; however, following the translation conventions in the field of physics, this book collectively refers to them as particles [36].

Quantum walks are divided into discrete-time quantum walks and continuous-time quantum walks. The former model mainly relies on coin operators and shift operators to provide evolutionary dynamics for the particle (some discrete-time quantum walks are coinless [37,38]), while the latter model relies on the Schrödinger equation to provide evolutionary dynamics that support the particle's transitions between different nodes on the graph. Regardless of the model, their study cannot be separated from specific graphs. For a quantum walk on any given graph, the particle must rely on the link relationships between nodes to move. Discrete-time quantum walks mostly construct shift operators by combining the standard bases of adjacent nodes, ensuring that the particle's trajectory is well defined. In continuous-time quantum walks, the adjacency matrix or Laplacian matrix of the graph is usually used to replace the Hamiltonian in the Schrödinger equation, ensuring that the particle moves by relying on the network's connectivity.

Chapter 2 provides a detailed introduction to the definition of quantum walks on existing computers and their simulation experiments. This part only demonstrates the operating principles of quantum walks on quantum devices. Taking the discrete-time quantum walk on a ring as an example [39], when the ring has a length of 16, the number of quantum circuit lines required for the nodes is 4, which is derived from, $\log_2 16$. A particle starting from any node on the ring has two possible walking directions, represented by $|0\rangle$ and $|1\rangle$, respectively. Therefore, the qubits in the quantum circuit diagram of the discrete-time quantum walk on the ring should be divided into two parts: four qubits in a superposition state used to represent the ring of length 16, and qubits used to indicate the particle's movement to its neighboring nodes. When the particle moves in the direction represented by $|0\rangle$, the quantum circuit executes the increment module "incr"; conversely, the circuit executes the decrement module "decr." In the quantum circuit shown in Figure 1.3, $|0\rangle$ and $|1\rangle$ act as control bits participating in the computation of the "incr" and "decr" models, thereby constructing the shift operator S of the quantum walk on the ring, where $S = \text{incr} \otimes |1\rangle + \text{decr} \otimes |0\rangle$. Since the ring is a closed shape, the circuit in Figure 1.3b can be executed cyclically.

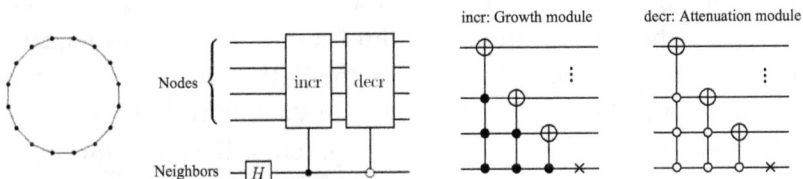

FIGURE 1.3   Quantum circuit diagram of a quantum walk on a ring [39].

1.2.3 HHL Algorithm

Solving systems of linear equations is a critical step in tasks such as eigen-value decomposition and classification in machine learning. Compared to classical non-quantum algorithms for solving linear systems, the HHL quantum algorithm achieves an exponential improvement in time complexity [3], which has strongly propelled the development of quantum machine learning.

Suppose the coefficient matrix of the equation to be solved is an $N \times N$ Hermitian matrix A, and b represents a unit vector. The goal of the HHL quantum algorithm is to find the vector x that satisfies $Ax = b$. Since a Hermitian matrix is a symmetric matrix, when matrix A is nonsymmetric, it can be transformed into a Hermitian matrix by expanding its dimension, while still satisfying the form $Ax = b$. The transformation method is:

$$\tilde{A} = \begin{pmatrix} 0 & A \\ A^{\dagger} & 0 \end{pmatrix}, \tag{1.32}$$

where $A^{\dagger}$ is the conjugate transpose of A. At this point, $\tilde{A}$ becomes a symmetric matrix, and $Ax = b$ is transformed into:

$$\tilde{A}y = \begin{pmatrix} b \\ 0 \end{pmatrix}. \tag{1.33}$$

Under the form of Eq. (1.33), for a non-Hermitian coefficient matrix A, the goal of the HHL quantum algorithm is to use $\tilde{A}$ to find the target solution y, which can be constructed as follows:

$$y = \begin{pmatrix} 0 \\ x \end{pmatrix}. \tag{1.34}$$

Considering the case where A is a Hermitian matrix, the solution approach of the HHL quantum algorithm is $x = A^{-1}b$. To obtain the quantum solution corresponding to the vector x, the HHL algorithm first expresses the matrix A and the vector b as quantum states in the form of standard basis states multiplied by probability amplitudes. The quantum state corresponding to vector b is defined as $|b\rangle = \sum_{i=1}^{N} b_i |i\rangle$; the matrix A can be decomposed into a combination of its eigenvalues and eigenvectors, that is,

$A = \sum_{j=0}^{N-1} \lambda_j |\mu_j\rangle\langle\mu_j|$, where $|\mu_j\rangle$ represents the eigenvectors of matrix A. Therefore, the inverse of matrix A is defined as $A^{-1} = \sum_{j=0}^{N-1} \lambda_j^{-1} |\mu_j\rangle\langle\mu_j|$. Based on the above approach solving for $x = A^{-1}b$, the target solution $|x\rangle$ of the HHL quantum algorithm is given as

$$|x\rangle = A^{-1}|b\rangle = \sum_{j=0}^{N-1} \lambda_j^{-1} b_j |\mu_j\rangle. \tag{1.35}$$

When executed on a quantum computer, the HHL quantum algorithm comprises three modules, including quantum phase estimation, rotation, and inverse quantum phase estimation. The corresponding quantum circuit can be found in Figure 1.4. In the figure, $FT^{\dagger}$ represents the conjugate transpose of the Fourier transform operator; operator R corresponds to a rotation gate; $H^{\otimes n}$ denotes the superposition of n qubits; and $U^{\dagger}$ indicates the conjugate transpose of U. In the quantum phase estimation module of Figure 1.4, the matrix A is expressed in the form of a Hamiltonian, denoted as $U = e^{iAt}$. The purpose of phase estimation is to obtain the eigenvalues λ_j of matrix A; that is, the eigenvalues λ_j are the results of the quantum phase estimation module. This result is employed as a control bit in the rotation module, adjusting the probability amplitudes of the eigenvalues through preset rotation angles. This process requires introducing a constant C and constructing a quantum state based on ancillary qubits:

$$\sum_{j=1}^{N} \left(\sqrt{1 - \frac{C}{\lambda_j^2}} |0\rangle + \frac{C}{\lambda_j} |1\rangle \right) \beta_j |\lambda_j\rangle |\mu_j\rangle. \tag{1.36}$$

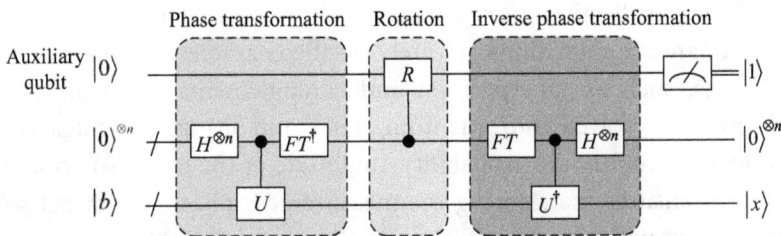

FIGURE 1.4   Quantum circuit diagram of the HHL quantum algorithm.

Finally, the inverse phase estimation operation is performed, and the ancillary qubits are measured. If the measurement result equals 1, the solution $|x\rangle^*$ is output; otherwise, recalculation is performed. The solution $|x\rangle^*$ obtained through the quantum circuit has a proportional relationship with the actual target solution $|x\rangle$, satisfying $|x\rangle^* \propto |x\rangle$.

It is worth mentioning that the Hamiltonian simulation stage of the HHL quantum algorithm—specifically, the process of constructing the operator U using e^{iAt}—can be implemented through quantum walk models [40]. This implementation can be directly applied to the phase estimation module of the HHL quantum algorithm [41]. The HHL quantum algorithm has become a favorite in the field of quantum machine learning [6], and many quantum versions of machine learning algorithms rely on HHL implementations, such as the quantum support vector machine [23], the quantum version of least squares fitting [42], quantum principal component analysis [43], and so on. In 2020, Origin Quantum Computing Technology Co., Ltd. in Hefei, China, utilized the HHL quantum algorithm based on the spring ranking model [44] to identify key nodes in complex networks [45], where the spring ranking model calculates the importance of nodes in directed networks through solutions of linear equations. This achievement provides a new solution approach for the application of quantum algorithms in complex networks.

This section only introduces a few representative quantum algorithms that are highly related to quantum walks. There are many other algorithms that can be considered groundbreaking work in quantum computing, such as Shor's integer factorization algorithm [1], the Deutsch-Jozsa algorithm [46], adiabatic quantum computing [26], and quantum simulated annealing algorithms [47]. Interested readers can access the online website "Quantum Algorithm Zoo" [48] for more information.

1.2.4 Relationship between Quantum and Non-Quantum Algorithms

As outlined in Section 1.2.1, it can be observed that the main design idea of quantum algorithms is to rely on the characteristics of quantum mechanics, such as superposition and entanglement. By encoding the problem to be solved into quantum states and using well-constructed operators to amplify the probability amplitude of the target solution, the algorithm ensures that, during the measurement phase, the target solution is output with a very high (close to 1) probability. Analyzing the relationships between quantum algorithms and classical algorithms helps to divide quantum algorithms into modules, improving the readability of the quantum algorithm process and clarifying the design ideas of quantum walk algorithms in environments like complex networks. First, quantum algorithms encompass some features of classical algorithms, including input, output, and finiteness. In this context, qubits in a quantum algorithm serve as the smallest units of input information, while the quantum measurement stage represents the output of the quantum algorithm. The finite evolutions in a quantum algorithm correspond to the finiteness seen in classical algorithms. Second, the evolution process in quantum algorithms aligns with the core processing procedures of classical algorithms, but in quantum algorithms, it is manifested solely through computational forms involving matrix multiplication.

In addition, starting from the basic postulates of quantum mechanics, we can consider the connections between quantum algorithms and classical algorithms. The state space defines the length of state vectors in quantum algorithms, which reflects the concept of space complexity in classical algorithms. Space complexity is often neglected in classical algorithms, partly because many algorithms do not involve the consumption of large amounts of storage space, and partly because emerging technologies represented by machine learning and deep learning can process data in batches, alleviating the problem of not being able to read large-scale data all at once. However, since matrices and operations between matrices serve as the core expressive forms in quantum algorithms, their space complexity cannot be ignored. Unitary evolution corresponds to the core steps of the algorithm in non-quantum algorithms. For instance, in the quantum genetic algorithm [14], using rotation gates can achieve random initialization of the population to obtain infinite possibilities for optimization; then, quantum NOT gate operations are used to mutate individuals in the population. The initialization and update strategies of the population are fundamental to swarm intelligence algorithms, directly affecting the algorithm's solution accuracy. In quantum genetic algorithms, the above core steps are all completed by matrix multiplication operations. Moreover, composite systems correspond to the inputs of classical algorithms and can be viewed computationally as tensor products between the states of each component. Quantum computing emphasizes treating the problem to be processed as an overall superposition, with the state vector serving as the input of the quantum algorithm. Finally, regarding the quantum measurement process, still using the quantum genetic algorithm as an example, this process is utilized in the algorithm to calculate the fitness of individuals in the population. It can act as an intermediate step to provide new heuristic information for the next population iteration optimization, and can also be output as the final result of the algorithm.

It is essential to discuss the robustness and feasibility of quantum algorithms, which rely on quantum computers to ensure these two characteristics. Factors such as the physical materials of quantum computers, implementation methods, error correction, and fault-tolerance capabilities cannot be obtained solely from the quantum algorithms themselves. For example, compared to superconducting quantum computers and semiconductor quantum computers, ion-trap quantum computers have a slight advantage in computational accuracy. Moreover, quantum measurement is the result of random collapse after finite evolutions and can only output the target solution with a probability close to 1. This leads to two thorny problems for quantum algorithms: first, due to the influence of random collapse during the measurement process, multiple measurements of the quantum circuit are required to obtain the desired result; second, the quantum circuits of quantum algorithms rely on quantum computers for implementation, and the performance of quantum algorithms may vary across different quantum computers. In this predicament, quantum walk algorithms are considered a universal computation model in quantum computing [29,49,50], as they are not constrained by the physical materials of quantum computers and can be used to design and simulate other quantum algorithms. This makes them particularly special among many quantum algorithms. Moreover, since quantum walks are extended from classical random walks, their simulation on existing von Neumann architecture computers does not require reliance on quantum computing cloud platforms or quantum computing toolkits of specific programming languages (such as Python's Qiskit toolkit). Simulation results can be obtained on existing computers through matrix operations. If we consider only the simulation of quantum walks on existing computers, quantum walk algorithms can liberate themselves from the constraints imposed by quantum devices and possess all the factors of classical (non-quantum) algorithms, including robustness.

This book will utilize quantum walks to mine and characterize meaningful structural information in complex networks, providing valuable references and guidance for applications such as viral marketing in social networks, friend recommendation, and protein functional module mining.

1.3 INTRODUCTION TO LOW-DIMENSIONAL QUANTUM WALKS

In 1993, as an extension of classical random walks into the quantum realm, research results on discrete-time quantum walks were first published in the journal *Physical Review A* [34]. Although 88 years had elapsed since the proposal of classical random walks in 1905 [51] and the introduction of discrete-time quantum walks, quantum walks rapidly began to play an irreplaceable role in areas such as universal computation models [28], marked point search [21], graph neural networks [52], and communication security [31]. This section introduces representative applications of quantum walks achieved in recent years from the perspectives of information security and spatial search. The implementation of quantum walks relies on the topological structure of graphs; therefore, their application domains differ based on different graph structures. Existing research on quantum walks generally focuses on low-dimensional regular graphs, such as lines, rings, two-dimensional lattices, and regular graphs. Hence, this section only introduces the applications of quantum walks in low-dimensional settings.

1.3.1 Low-Dimensional Quantum Walks in Information Security

When quantum walks occur on low-dimensional graphs, their primary applications are in designing communication protocols and pseudorandom number (key) generators. In such known applications, quantum walks specifically refer to discrete-time quantum walks (also known as coin quantum walks). The probability distribution characteristics of quantum walks are highly related to the settings of the initial point and initial state, and they exhibit chaotic behavior. Building on this idea, reference [53] proposes a cascaded quantum walk method with a chaotic system, which generates random sequences based on probability distribution results and is applied to image encryption. Yang et al. designed a random number generator based on two-particle quantum walks [54], further enhancing the encryption effect of existing single-particle quantum walks. A year later, the team proposed a construction scheme for hash functions using controlled two-particle quantum walks for image encryption and proved its effectiveness [55]. Similarly, a substitution box method for image steganography was designed based on two-particle quantum walks in [56]. The aforementioned keys are all generated based on two-particle quantum walks, and their corresponding Hilbert spaces lead to an expansion of operator dimensions due to the execution of tensor product calculations. This requires consuming a significant amount of storage space, thereby limiting the generation of random numbers. Consequently, reference [57] developed a single-particle quantum walk encryption algorithm and applied this method to 5G Internet of Things to achieve secure transmission.

The aforementioned applications are limited to pseudorandom number generation using quantum walks. However, quantum walks can also serve as designs for communication protocols in the field of information security. Wang et al. proposed a generalized communication protocol based on quantum walks with alternating two coins [58]. This protocol achieves quantum teleportation of an unknown qubit state by utilizing a two-step quantum walk on a line and a quantum walk on a ring of length 4. To accommodate the transmission of arbitrary states on high-dimensional regular graphs, Shang et al. further improved this work in 2019 [59]. Specifically, they replaced the original protocol's d^2 dimensional quantum measurement with two d dimensional measurements, simplified the original quantum walk, and provided a general model and quantum circuit diagram for two-coin quantum walks.

Quantum walks can also be applied for quantum signatures. Similar to classical signature problems, quantum signatures are also categorized into two-party quantum signatures and arbitrated quantum signatures. In arbitrated quantum signatures, an arbiter is introduced in addition to the sender and receiver to verify the information. The arbiter attaches a time-stamp to the verified message to ensure reliable communication between the sender and receiver. Barnum et al. pointed out that absolutely secure two-party quantum signature protocols do not exist [60], and from a practical perspective, arbitrated signature schemes involving a trusted third party are more practical [61]. Therefore, most research on quantum signatures focuses on arbitrated quantum signature schemes. For instance, Feng et al. introduced an arbitrated quantum signature scheme with quantum walk-based teleportation [62]. In their scheme, the entangled states are naturally generated during the signing stage through quantum walks, eliminating the need for prior preparation. Security analysis shows that the signatures in this scheme are non-repudiable. In a like manner, they proposed an arbitrated quantum signature algorithm using quantum walks on regular graphs, which also does not require pre-preparation of entanglement sources [63]. Additionally, Li et al. designed a quantum information-splitting scheme based on multi-coin state quantum walks

[64]. This scheme avoids the need to prepare entangled states in advance or measure the degree of entanglement, thereby reducing the resource consumption of quantum network communication.

1.3.2 Low-Dimensional Quantum Walks in Spatial Search

Searching for a set of marked nodes on a graph is known as the spatial search problem. Around the year 2000, numerous definitive conclusions had been formed regarding the application of quantum walks to this problem. For discrete-time quantum walks, the time required to find a single marked node on any ergodic reversible Markov chain is only the square root of the time needed by classical random walks [65]. Nevertheless, in the spatial search problem based on continuous-time quantum walks, accelerated search can be achieved only on specific types of graphs [28]. For example, on two-dimensional lattices, three-dimensional lattices, and four-dimensional lattices, the time complexities of spatial search algorithms based on continuous-time quantum walks are $O(N), O(N^{5/6})$, and $O(\sqrt{N}\log N)$, respectively, where N represents the total number of nodes in the graph; whereas when the lattice dimension is greater than 5, the time complexity becomes $O(\sqrt{N})$ [28].

In 2004, Ambainis et al. proposed a coined quantum walk—the AKR (Ambainis, Kempe, Rivosh) algorithm [66]—which classifies marked and unmarked vertices on a graph and represents them using different coin operators. Analysis shows that the AKR algorithm exhibits a time complexity of $O(\sqrt{N}\log N)$ for searching marked vertices on a two-dimensional lattice. When the lattice dimension exceeds 3, the AKR algorithm can achieve square-root speedup. Another algorithm that realizes accelerated search is the well-known Szegedy quantum walk algorithm [37]. The main idea of this algorithm is to map the link relationships among nodes in the graph to a bipartite graph, independently represent the two parts of the bipartite graph as quantum states, and then compose them to form an evolution operator for search. On one-dimensional, two-dimensional, and lattices with dimensions greater than or equal to 3, the time complexities of spatial search using the Szegedy quantum walk are $O(N), O(\sqrt{N}\log N)$, and $O(N^{d/2})$, respectively, where d represents the lattice dimension.

In fact, quantum algorithms fundamentally differ from classical (non-quantum) algorithms. When solving spatial search problems using quantum walk algorithms, although the time complexity is lower, the measurement probability corresponding to the marked vertex may not be the highest [65].

In other words, in quantum algorithms, detecting the existence of a marked vertex on the graph does not necessarily mean that the algorithm can output that marked vertex with a very high measurement probability. As an illustration, in the spatial search results of the Szegedy quantum walk on a two-dimensional lattice, the hitting time (time complexity) to the marked vertex is $O(\sqrt{N}\log N)$, but the measurement probability of the marked vertex is $1/\log N$. Therefore, the search for marked vertices by this quantum walk on a two-dimensional lattice is limited to detecting their existence and cannot find them with a very high probability. In 2011, Magniez et al., based on the design concepts of reversible Markov chains and bipartite graphs, utilized the phase estimation method to recursively amplify the probability amplitude of the marked vertex [67], enabling it to be measured with a very high probability. This was a successful step toward solving the problem of low measurement probability. In the study of [67], although the measurement probability of the marked vertex was greatly improved, it weakened the speedup advantage of the quantum walk in terms of hitting time. In 2012, Magniez et al. redefined a parametric Monte Carlo hitting time evaluation formula [68]. Their conclusion pointed out that quantum walks can achieve quadratic speedup in search time while also finding the marked vertex with a very high probability.

Although the breakthroughs achieved by quantum walks in the spatial search problem have been arduous, related results have continuously extended their applications in other fields. A highly representative work is the element distinctness problem. Element distinctness, also known as element discrimination, aims to determine whether there exist two identical elements among N elements. In the quantum algorithm implemented by Buhrman et al. [69], the element distinctness problem is solved with a time complexity of $O(N^{3/4})$. Ambainis further improved the solution efficiency with an algorithm based on quantum walks [70], allowing the element distinctness problem to be solved with a time complexity of $O(N^{2/3})$. The aforementioned quantum walk results by Ambainis have also been extended to applications such as triangle finding in graphs [30] and matrix product verification [71]. Because the theoretical proofs involved in the applications of quantum walks in spatial search are particularly intricate, interested readers can refer to the original papers mentioned above or to works represented by Wong et al. [72–74] and Chakraborty et al. [75].

Information security and spatial search are only two highly representative branches among the application domains of low-dimensional quantum walks. If the graphs on which quantum walks rely are not low-dimensional regular graphs—for example, irregular complex networks—then quantum walk models will have generalization capabilities, and their application domains will be further expanded, because regular graphs can be regarded as special cases of complex networks. From the applications of quantum walks in spatial search, it is apparent that existing works are not satisfied with the advantages presented by quantum walks on regular and special graphs [76–78]. Researchers have begun to explore the relationship between network statistical properties and search efficiency. In 2010, Berry et al. investigated the relationship between the search probability of marked nodes on a graph and node centrality when the spatial search problem is addressed on arbitrary graphs [79]. In 2016, Chakraborty et al. highlighted that when the connection probability of random complex networks meets specific conditions, continuous-time quantum walks are the optimal solution for the spatial search problem [75]. These two studies reflect that, influenced by the practical demands of real-world applications, quantum walks on complex networks (arbitrary graphs) are gradually emerging. Against the backdrop of this rising interest in quantum walks on complex networks, this book explores the design ideas of quantum walk algorithms on complex networks, designs quantum walk algorithms, and introduces their applications in network node mining, link prediction, community detection, and network representation learning.

1.4 STRUCTURE OF THE BOOK

The entire book is divided into two parts. Chapters 1 and 2 lay the theoretical foundation, while Chapters 3–6 focus on the applications of quantum walk in information mining from complex network structures. Figure 1.5 illustrate the organizational structure of the book. Specifically, Chapter 2, building on the research of quantum walk on regular graphs, presents the current state of research on quantum walk in complex networks and discusses the design concepts and general framework of quantum walk algorithms on complex networks. The minimal components of a complex network are nodes and links, and mining meaningful nodes and links within the network is a fundamental topic in the field of complex networks. Chapter 3 introduces the applications of both discrete-time quantum walk and continuous-time quantum walk in mining key nodes within complex networks. Chapter 4 focuses on two applications: the identification of critical links and link prediction, showcasing the application of quantum walk

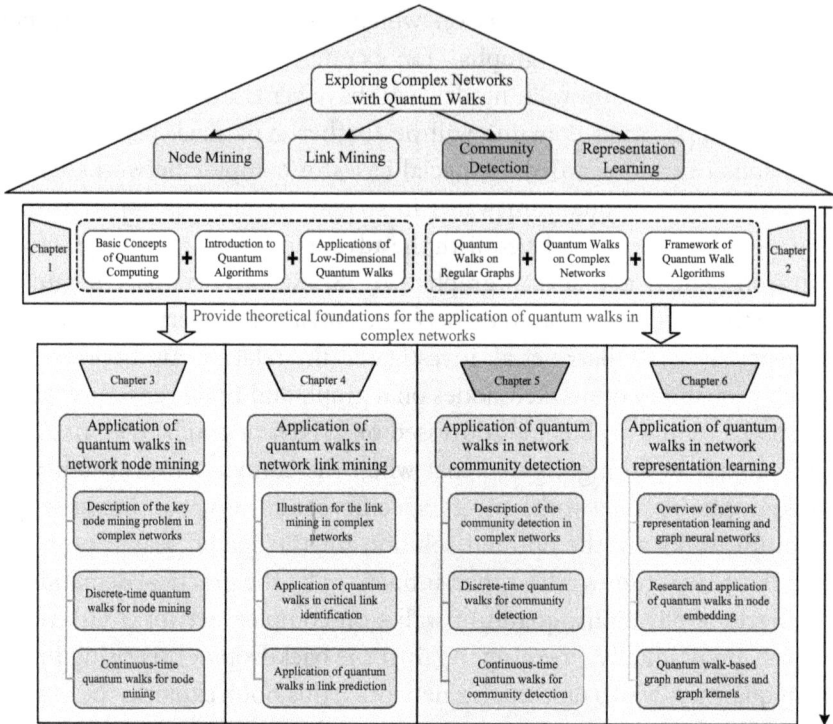

FIGURE 1.5 The organizational structure of this book.

algorithms in link mining. Furthermore, the scope of the mining targets can be extended to subgraph structures composed of nodes and links. Consequently, Chapter 5 discusses the application of quantum walk in community detection within complex networks, where communities are defined as subgraph structures of particular significance. Finally, Chapter 6 presents the applications of quantum walk algorithms in network representation learning and graph neural networks, including node similarity calculations, network classification, and graph isomorphism, concluding with an analysis of future research directions for quantum walk in network representation learning.

CHAPTER 2

Foundational Theory of Quantum Walks

Q UANTUM WALKS, ALTHOUGH EXTENDING from classical random walks, exhibit characteristics that are fundamentally different from those of classical random walks. Comparing the traversal process to the traversal of nodes on a graph, the classical random walk follows a depth-first search, where the particle starts from the current node and visits one of its neighboring nodes at each step. In contrast, the traversal process in quantum walks resembles a breadth-first search, where the particle departs from the current node and visits all neighboring nodes in the nearest layer at each step. For quantum walks, when the standard basis states of the network nodes are in a superposition state, it is equivalent to the particle creating a "clone" at each node, or it can be understood as having a particle at each node. In this case, the walk exhibits the characteristics of superposition and the concept of parallelism. The effects of superposition and interference result in the probability distribution of quantum walks differing significantly from that of classical random walks, and they also inject new vitality into the study of quantum walks. For instance, the superposition and interference effects of quantum walks can amplify the probability amplitude of similar nodes in isomorphic graphs [52]. Compared to classical (non-quantum) algorithms, quantum walks can solve spatial search problems with a square root time complexity [75], and interference effects in random quantum walks can be used to generate pseudorandom numbers by treating them as chaotic systems [80]. The interference

DOI: 10.1201/9781003683902-3

effect, a byproduct of the superposition state, is a common characteristic shared by quantum algorithms. If quantum walks are to become effective algorithms for solving complex network structure mining problems, the interference effect is just one of the factors that need to be considered in designing quantum walk algorithms. Other aspects, such as the step length of quantum walks and the construction of evolution operators, are also crucial in designing quantum walk algorithms. Therefore, unlike existing literature, this chapter not only focuses on the measurement results and characteristics (e.g., central limit and localization) caused by the interference effect of quantum walks but also emphasizes the design and conceptualization of quantum walk algorithms on complex networks. Currently, quantum walks are mainly studied on low-dimensional regular graphs, such as one-dimensional lines, rings, and two-dimensional lattices, while research on quantum walks on complex networks remains insufficient.

This chapter serves as the theoretical foundation of quantum walk algorithms throughout the book, extending from quantum walks on regular graphs to those on complex networks and providing a general framework for quantum walk algorithms on complex networks.

2.1 QUANTUM WALKS ON REGULAR GRAPHS

Research on quantum walks in low-dimensional regular graphs forms the foundational theory of quantum walks and acts as the core framework for guiding their application in high-dimensional and irregular graphs [80]. Currently, the main foundational theories of quantum walks are covered in *Quantum Walks and Search Algorithms* by Portugal [21] and *Physical Implementation of Quantum Walks* by Wang et al. [36]. The former focuses on explaining quantum walks in quantum computing and their related search algorithms, while the latter emphasizes the physical implementation of quantum walks, particularly in optical experiments. In this section, we take the one-dimensional line and the two-dimensional lattice as examples to study the characteristics of the measurement results of discrete-time and continuous-time quantum walks. This study aims to accumulate theoretical evidence and provide design references for the development of quantum walk algorithms on complex networks.

2.1.1 Low-Dimensional Discrete-Time Quantum Walks

2.1.1.1 One-Dimensional Discrete-Time Quantum Walks

The definition and study of discrete-time quantum walks begin with the one-dimensional line [80]. The characteristic of this walk is that after tossing a coin, the direction of the particle's movement is determined by whether the coin lands on heads or tails. Suppose the walk takes place on a line L. In a one-dimensional discrete-time quantum walk, except for the nodes at the endpoints, a particle starting from any node has two possible directions to choose from—left or right—typically described using the orthogonal basis $|0\rangle$ and $|1\rangle$. If the length of line L is n, and each point on the line is represented by the orthogonal basis $|j\rangle$ of length n, the Hilbert space of the line L is composed through the tensor product operation of the orthogonal basis representing the walk direction and the orthogonal basis for each point, expressed as $\mathcal{H} = \mathcal{H}^2 \otimes \mathcal{H}^n$. Consequently, the dimension of the Hilbert space on line L is determined to be $2n$. In other words, on the line L, the dimensions of the state vectors and the evolution operators are both equal to $2n$. Furthermore, an initial state vector of length $2n$ can be defined as follows:

$$|\psi(0)\rangle = \sum_{1}^{n} \alpha_j(0)|j\rangle, \qquad (2.1)$$

where $\alpha_j(0)$ indicates the probability amplitude in relation to point j at the initial moment. The amplitudes of all points on the line L satisfy the following condition:

$$\sum_{1}^{n} |\alpha_j|^2 = 1. \qquad (2.2)$$

The probability amplitude constrains the movement of the particle on the line L. On one hand, the probability amplitude can specify that the particle starts from a particular point (or multiple points) on the line. On the other hand, the probability amplitude directly affects the probability distribution pertaining to the measurement results. To facilitate observation of the measurement results, this section sets the initial probability amplitude at the midpoint of the line to be 1, while the initial probability amplitudes of all other points are set to 0. In other words, the particle is specified to start from the center of the line L.

Subsequently, the evolution operator is employed to implement the process of the particle jumping from its current position to its neighboring point. This process involves tossing a coin and determining whether the coin shows heads or tails to decide the particle's walking direction, that is, left $|0\rangle$ or right $|1\rangle$. The shift operator is then used to execute the movement. The coin operator is the sum of all walking directions, specifically the total of the outer products of the $|0\rangle$ and $|1\rangle$ states, as outlined in Eq. (1.11) and denoted here as H. Moreover, the shift operator, which moves the particle based on the walking direction determined by the coin toss, is denoted as S_L and can be expressed as

$$S_L = |0\rangle\langle 0| \otimes \sum_j |j+1\rangle\langle j| + |1\rangle\langle 1| \otimes \sum_j |j-1\rangle\langle j|, \qquad (2.3)$$

where $|j+1\rangle$ indicates the particle moving from point j to its left neighboring point, and similarly, $|j-1\rangle$ represents the movement to the right neighboring point. The evolution operator U_L is composed of the shift operator S_L and the coin operator H as follows:

$$U_L = S_L \cdot \left(H \otimes \hat{I} \right), \qquad (2.4)$$

where the symbol $\hat{I}$ denotes an identity matrix of size n. Although the walking process involves tossing the coin first and then moving, Eq. (2.4) writes the operator $H \otimes \hat{I}$, representing the coin-tossing process, after the shift operator S_L, which represents the movement process. This order follows the conventions of quantum computing expressions [20], with a detailed explanation provided in Section 1.1.3. Based on the state vector defined in Eq. (2.1) and the evolution operator formulated in Eq. (2.4), the t-step walk of the particle on the line L can be expressed as

$$|\psi(t)\rangle = U_L^t |\psi(0)\rangle, \qquad (2.5)$$

where U_L^t represents the application of the evolution operator t times, denoted as $U_L^t = (U_L)^t$.

Finally, the probability distribution of the particle staying at each point is obtained through the quantum measurement process. Considering the measurement at point j as an example, after t steps of the walk, the probability of the particle staying at point j is defined as

$$P_j(t) = \left| \langle j | \psi(t) \rangle \right|^2 = \left| \langle j | U_L^t | \psi(0) \rangle \right|^2. \qquad (2.6)$$

Obviously, the formulation of Eq. (2.6) is a component of the calculation presented in Eq. (2.2). Therefore,

$$\sum_{j=1}^{n} P_j(t) = 1. \qquad (2.7)$$

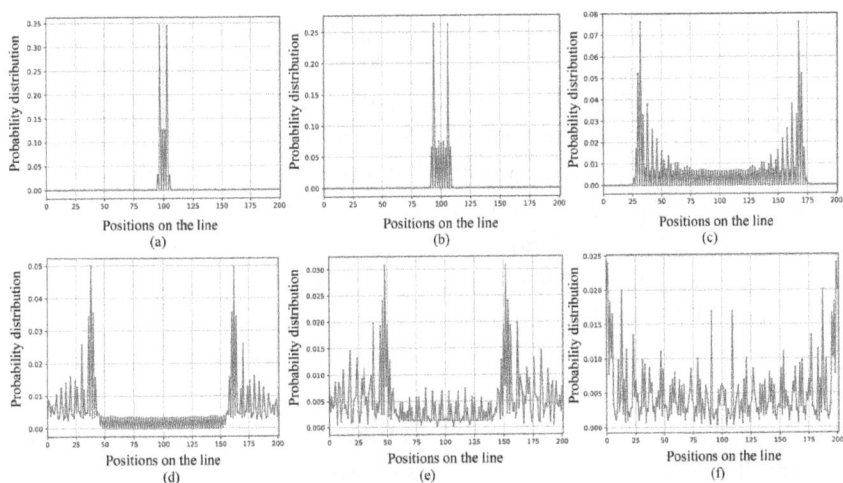

FIGURE 2.1   Distributions of a quantum walk on a line for different step lengths. (a) Evolution steps for $t=5$. (b) Evolution steps for $t=10$. (c) Evolution steps for $t=100$. (d) Evolution steps for $t=201$. (e) Evolution steps for $t=500$. (f) Evolution steps for $t=1,000$.

In this experiment, the probability of the particle staying at different points is ultimately presented in the form of a probability distribution. According to the above definition, suppose the discrete-time quantum walk takes place on a line of length 201, with the center of the line being the initial starting position for the walk. Figure 2.1 shows the probability distribution results measured for the quantum walk on the one-dimensional line under various step lengths.

Regarding the one-dimensional discrete-time quantum walk, the analysis and conclusions are as follows. When the walk step length is less than or equal to the length of the line ($t \leq n$), the characteristics of the measured probability distribution are related to whether the step length is odd or even. According to Figure 2.1a, the probability values at the positions corresponding to the scales 99 and 101 on the line are equal to 0, while the probability measured at the initial position (scale 100) is greater than 0. Furthermore, by comparing Figure 2.1a and b, when the walk step lengths are 5 and 10, respectively, the probability value observed at the initial position is greater than 0 for the step length of 5 and equal to 0 for the step length of 10. This indicates that the probability distribution measured for the one-dimensional discrete-time quantum walk is influenced by whether the step length is odd or even. Additionally, as shown in Figure 2.1e and f, when the walk step length is much greater than the length of the line, the probability at any point on the line is greater than 0.

The length of the walk step determines the distance covered and the maximum value in the measured probability distribution. According to Eq. (2.7), since the quantum walk satisfies the condition of unitary transformation, the sum of the squares of the probability amplitudes must always equal 1. The step length influences the distance the particle travels on the line, and this distance, in turn, affects how much probability is distributed across the points visited. For instance, in Figure 2.1a and c, as the step length increases from 5 to 100, the walking distance extends from near to far, and the maximum values of their probability distributions are approximately 0.35 and less than 0.08, respectively. This suggests that when the step length is less than or equal to the line length ($t \leq n$), the walking distance is directly proportional to the step length— implying that the greater the step length, the farther the walking distance. Consequently, the average occupancy rate of all nodes decreases. For more conclusions regarding the numerical characteristics of probability distributions in low-dimensional discrete-time quantum walks, refer to studies on the occupancy rate of quantum walk measurement results [81].

When the walk step lengths are 500 and 10^3, respectively, there is no apparent correlation between the step length and the measured probability distribution. To further explore the relationship between them, the real and imaginary parts of the eigenvalues of the evolution operator under diverse step lengths are visualized as scatter plots on a unit complex plane, revealing their intrinsic connections. The results are shown in Figure 2.2. Each scatter point in Figure 2.2 is determined by the real and imaginary parts of the corresponding eigenvalue. Comparing the three subfigures (a),

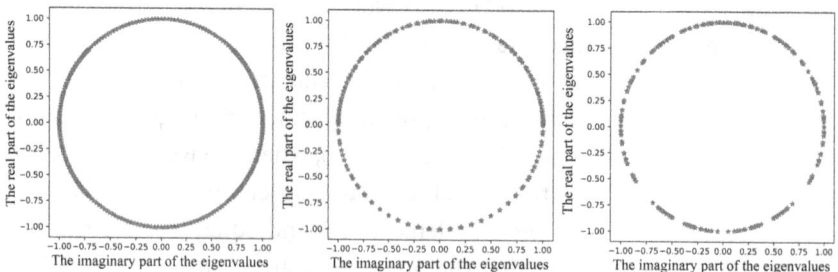

FIGURE 2.2   Distributions of eigenvalues of the evolution operator on the unit complex plane.

(b), and (c) in Figure 2.2, it is evident that as the step length increases, the distribution of the real and imaginary parts of the eigenvalues becomes increasingly uneven. In contrast to Figure 2.1, the regularity of the measured probability distribution deteriorates with increasing step length. It can be inferred that in a one-dimensional discrete-time quantum walk, the distribution of the real and imaginary parts of the eigenvalues is highly correlated with the degree of chaos in the measurement results. This is one of the reasons why many studies consider the measurement results of quantum walks to exhibit chaotic behavior and use them for designing random number generators [53,82,83].

2.1.1.2 Two-Dimensional Discrete-Time Quantum Walks

In this section, we will construct two-dimensional discrete-time quantum walks driven by three different coin operators and study the characteristics of the probability distributions measured for these three types of two-dimensional quantum walks.

In a two-dimensional lattice, which, like the line, is a regular graph, it is important to note that the two dimensions represent the horizontal and vertical directions, with the number of points in each dimension referred to as the length. In a two-dimensional lattice, except for the edge points, the number of possible walking directions for a particle at any point is 4. Therefore, the states $|0\rangle$ and $|1\rangle$ used in the one-dimensional discrete-time quantum walk are insufficient to describe the possible walking directions in the two-dimensional lattice. In this case, $|0\rangle$ and $|1\rangle$ need to be extended using the tensor product operation to increase their dimensions and expand their meaning in terms of walk direction selection. For example, the walking direction to the left among the 4 possible directions can be represented as

$$|0\rangle \otimes |0\rangle = |00\rangle = (1,0)^{\mathrm{T}} \otimes (1,0)^{\mathrm{T}} = (1,0,0,0)^{\mathrm{T}}. \qquad (2.8)$$

In the same way, the remaining three standard bases for the walking directions can be obtained, denoted as $|01\rangle$, $|10\rangle$, and $|11\rangle$, with their forms similar to Eq. (2.8), which will not be repeated here. In the two-dimensional discrete-time quantum walk, the two-dimensional lattice under study is abstracted into a quantum state. Let $|\psi_{2d}(0)\rangle$ be the quantum state at the initial moment. The points on the two-dimensional lattice with resemble two-dimensional coordinates; thus, any point j on the lattice is determined by two parameters, that is, x and y. The standard basis $|x,y\rangle$ signifies the particle's current position on the two-dimensional lattice. The symbol $|j_x, j_y\rangle$ denotes the standard basis corresponding to the possible walking direction at point j, where $j_x, j_y \in \{0,1\}$ indicate the direction parameters. Let the symbol $\alpha_{j_x, j_y, x, y}(t)$ express the probability amplitude at the point $|x, y\rangle$ after t steps, where the walk direction is given by $|j_x, j_y\rangle$. Hence, the state vector $|\psi_{2d}(t)\rangle$ can be described as

$$|\psi_{2d}(t)\rangle = \sum_{j_x, j_y=0}^{1} \sum_{x, y=0}^{n} \alpha_{j_x, j_y, x, y}(t) |j_x, j_y\rangle |x, y\rangle. \tag{2.9}$$

In the expression, the probability amplitudes satisfy the condition that the sum of their squared magnitudes equals 1, that is,

$$\sum_{j_x, j_y=0}^{1} \sum_{x, y=0}^{n} \left|\alpha_{j_x, j_y, x, y}(t)\right|^2 = 1. \tag{2.10}$$

The calculation process for a particle jumping between different positions on a one-dimensional line is similar to that for a particle on a two-dimensional lattice. The particle's movement still relies on an evolution operator composed of a coin operator and a shift operator. This section mainly discusses the effects of three types of coin operators on the probability distribution of quantum walks: the Fourier coin operator, the Hadamard coin operator, and the Grover coin operator. The definitions of these operators can be found in Eqs. (1.14), (1.11), and (1.15), respectively. Although three different coin operators are used for comparison in this section, the movement of the particle on the two-dimensional lattice requires only one shift operator, meaning that the discrete-time quantum walks driven by the three different coin operators share the same shift operator. The definition of the shift operator for two-dimensional discrete-time quantum walks is as follows:

$$S_{2d} |j_x, j_y\rangle |x, y\rangle = |j_x, j_y\rangle |x + (-1)^{j_x}, y + (-1)^{j_y}\rangle. \tag{2.11}$$

Since the particle on the two-dimensional lattice includes four possible walking directions, $|x + (-1)^{j_x}, y + (-1)^{j_y}\rangle$ represents four distinct movement directions: $|x-1, y-1\rangle$, $|x-1, y+1\rangle$, $|x+1, y-1\rangle$, and $|x+1, y+1\rangle$. The three coin operators mentioned above are denoted as C_{coin}, where $\text{coin} = \{F, G, H\}$, with F, H, and G denoting the Fourier, Hadamard, and

Grover coin operators, respectively. Thus, the evolution operator for the two-dimensional discrete-time quantum walk can be defined as follows:

$$U_{2d} = S_{2d}(C_{\text{coin}} \otimes \hat{I}).$$ (2.12)

Based on the definition of the state vector in Eq. (2.9), a single step of the two-dimensional discrete-time quantum walk can be detailed below:

$$|\psi_{2d}(t+1)\rangle = \sum_{j_x,j_y=0}^{1} \sum_{x,y=0}^{n} \alpha_{j_x,j_y,x,y}(t) S_{2d}\left(C_{\text{coin}}|j_x,j_y\rangle|x,y\rangle\right)$$

$$= \sum_{j_x,j_y=0}^{1} \sum_{x,y=0}^{n} \alpha_{j_x,j_y,x,y}(t) C_{\text{coin}}|i_x,i_y\rangle|x+(-1)^{j_x}, y+(-1)^{j_y}\rangle.$$

(2.13)

Following the above definitions, the experiment for the two-dimensional discrete-time quantum walk is designed as follows: assume a two-dimensional lattice with a size of 41×41, the particle's initial walking position is at the center point $(0,0)$ of the lattice, with equal probability amplitudes assigned to the four walking directions (up, down, left, and right) from this point. According to Eq. (2.10), all four directions are set to 0.5, while the initial probability amplitudes at all other lattice points are set to 0. Taking the two-dimensional discrete-time quantum walk driven by the Hadamard coin operator as an example, the probability amplitude at the initial center point is expressed as follows:

$$\alpha_{0,0,0,0}(0) = \alpha_{0,1,0,0}(0) = \alpha_{1,1,0,0}(0) = \frac{1}{2}, \quad \alpha_{1,0,0,0}(0) = -\frac{1}{2}.$$ (2.14)

Figures 2.3 and 2.4 show the measurement results of the two-dimensional discrete-time quantum walk driven by three different coin operators when the walking steps are set to 10 and 20, respectively. In each subplot, the color bar on the right indicates the range of the measured probability values at each position on the two-dimensional lattice; the darker the color, the higher the measured probability value at that position, indicating that the particle is most likely to collapse (stay) at the corresponding position. It can be observed that when the walking steps are less than the lattice size, the probability distribution characteristics for steps 10 and 20 are similar. According to Figures 2.3 and 2.4, the probability distribution associated with the Fourier coin operator exhibits asymmetry along the diagonal direction, while the distributions relevant to the Grover and Hadamard coin operators display symmetry in all directions. In the experiments involving the Fourier and Grover coin operators, the higher probability values are located at the ends of the horizontal and vertical directions, while in the Hadamard coin operator's results, the higher probability values are distributed at the four corners. Additionally, it can be seen from Figures 2.3 and 2.4 that the Hadamard coin operator relates to the shortest walking distance, while the Grover coin operator aligns with the longest walking distance under the same initial conditions.

In Figures 2.3 and 2.4, the initial probability amplitudes at the starting point are set equally. Figure 2.5 depicts the experimental results for the

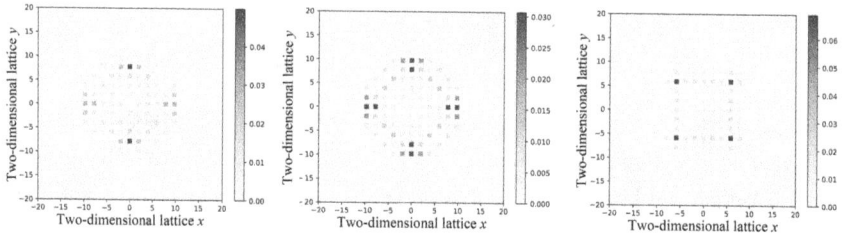

FIGURE 2.3   Distributions of a two-dimensional quantum walk with $t = 10$.

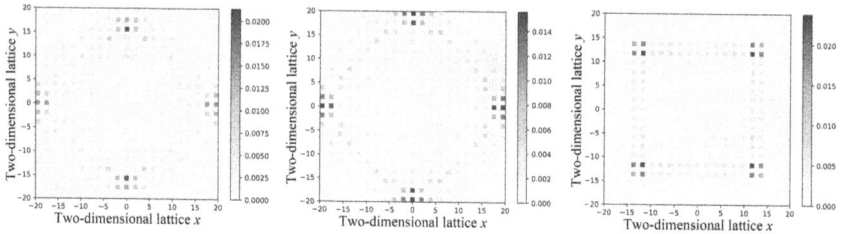

FIGURE 2.4   Distributions of a two-dimensional quantum walk with $t = 20$.

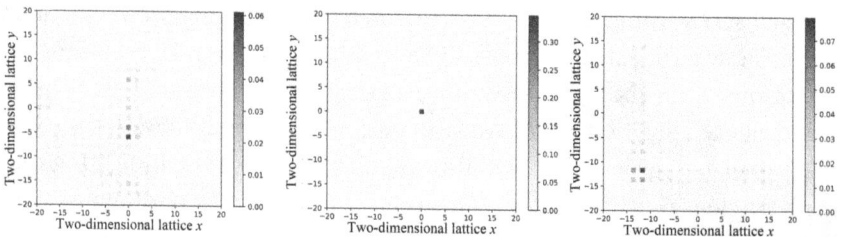

FIGURE 2.5   Biased distributions of a two-dimensional discrete-time quantum walk.

three types of two-dimensional discrete-time quantum walks with a walking step length of 20, where the initial probability amplitude at the starting point is set with a bias. The probability amplitude at the center point $(0,0)$ is defined as $\alpha_{0,0,0,0}(0)=1$, specifying only the walking direction corresponding to the coin state $|00\rangle$, while the initial probability amplitudes for all other walking directions are set to 0; the probability amplitudes for the other three directions from the center point are also set to 0, that is, $\alpha_{0,1,0,0}(0)=\alpha_{1,0,0,0}(0)=\alpha_{1,1,0,0}(0)=0$. In Figure 2.5, the color bar on the right side of each subplot indicates the range of the measured probability values at each position on the two-dimensional lattice, with darker colors representing higher measured probabilities. According to Figure 2.5, starting from the center point $(0,0)$ and choosing only the walking direction corresponding to the $|00\rangle$ state, the probability distributions for the three quantum walks are all biased, and the direction of bias is toward the lower left corner. Furthermore, in comparison with the experimental results in Figure 2.4, the walking distance characteristics of the three quantum walks exhibit significant changes. The quantum walks driven by the Grover coin operator now results in the shortest walking distance, while the one driven by the Fourier coin operator leads to the longest under identical initial conditions. Similarly, it can be inferred that under other forms of biased initial probability settings at the center point, the measurement results of the quantum walk will also exhibit biased distributions.

2.1.2 One-Dimensional Continuous-Time Quantum Walks

Unlike discrete-time quantum walks on lower-dimensional graphs, it is not essential to distinguish whether the graph relied upon by continuous-time quantum walks is low-dimensional, high-dimensional, regular, or irregular, as continuous-time quantum walks only require reading the adjacency matrix pertaining to the graph structure. In this section, the one-dimensional continuous-time quantum walk is a shorthand for the continuous-time quantum walk on a one-dimensional line, and the related definitions are still applicable to irregular graphs, such as those represented by complex networks.

Assuming that the continuous-time quantum walk occurs on a graph G, the Schrödinger equation provides the dynamics that drive the particle's transitions between different nodes. The Schrödinger equation can be expressed as follows:

$$i\frac{d}{dt}|\psi(t)\rangle = H|\psi(t)\rangle. \tag{2.15}$$

The purpose of Eq. (2.15) is to record the time evolution of the quantum state generated by the Hamiltonian, where the symbol H represents the Hamiltonian, which can be replaced by the adjacency matrix or the Laplacian matrix associated with the graph G. If the graph under study is a regular graph, then the effects produced by the adjacency matrix and the Laplacian matrix are equivalent [84]. In this section, the adjacency matrix pertaining to graph G is used. $|\psi(t)\rangle$ indicates the state vector of graph G, and similar to the state vector defined in discrete-time quantum walks, $|\psi(t)\rangle$ also consists of probability amplitudes α_j and the standard basis $|j\rangle$:

$$|\psi(t)\rangle = \sum_j \alpha_j(t)|j\rangle. \tag{2.16}$$

In the equation, j signifies any node in graph G. Solving the Schrödinger equation in Eq. (2.15) yields the evolution formula for the state vector:

$$|\psi(t)\rangle = e^{-iHt}|\psi(0)\rangle. \tag{2.17}$$

Let the adjacency matrix of graph G be A. By substituting the adjacency matrix for the Hamiltonian H in the Schrödinger equation of Eq. (2.17), the evolution formula can be written as follows:

$$|\psi(t)\rangle = e^{-iAt}|\psi(0)\rangle. \tag{2.18}$$

Finally, quantum measurement is used to observe the probability of the particle staying at each node, resulting in the probability distribution on graph G. The measurement method is as follows:

$$P_j(t) = |\alpha_j(t)|^2. \tag{2.19}$$

Analogous to the implications of Eqs. (2.2) and (2.10), the measurement results of the continuous-time quantum walk also satisfy the condition that the sum of probabilities equals 1.

In this section, taking a line of length 101 as an example, the center position of the line is chosen as the particle's starting point. Figure 2.6 shows the measurement results of the one-dimensional continuous-time quantum walk. It can be observed that when the value of t is much smaller than the length of the line, as t increases, the particle's walking distance

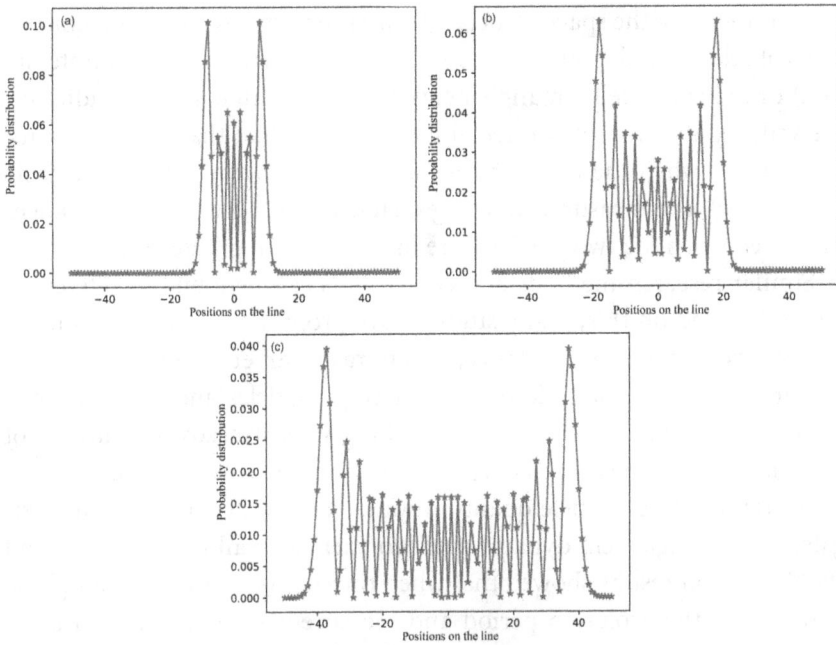

FIGURE 2.6 Distributions of a one-dimensional continuous-time quantum walk. (a) $t=5$. (b) $t=10$. (c) $t=20$.

on the line becomes farther, exhibiting the characteristic of a lower measurement probability at the starting point and higher measurement probabilities at the endpoints.

The experiments and discussions on discrete-time and continuous-time quantum walks on regular graphs presented above will provide strong support for the design and conceptualization of quantum walk algorithms on complex networks in subsequent sections. The design ideas for the relevant algorithms are detailed in Sections 2.2.3 and 2.3. To showcase the research results of quantum walks from multiple perspectives, the following section provides a brief introduction to the variant studies of quantum walks on regular graphs.

2.1.3 Variants of Quantum Walks on Regular Graphs

The study of variants in discrete-time quantum walks mainly focuses on the amplification of the number of particles, typically using two particles as an example, and most studies are based on a one-dimensional line. In the one-dimensional quantum walk of two particles, two identical particles are distinguished using bosons and fermions according to the symmetry or asymmetry of the space associated with the particles. The main conclusions of related studies are as follows: if the two particles interact (interaction) or are entangled (entanglement), under the same initial conditions, the walking distance of two fermions is greater than that of two bosons. In contrast, when the two particles are non-interacting (distinguishable or separate), the measured walking distance falls between the distances measured when the two particles are in an interacting state [85].

Similar research methods and conclusions can be found in references [86–88]. Based on this, many studies have proposed detailed hypotheses and verified them. For instance, Chandrashekar et al. hypothesized a scenario involving multiple non-interacting particles and found experimentally that when the number of walking steps exceeds the number of particles, the joint measurement results of the multi-particle system resemble those of single-particle quantum walks [89]. Rodriguez et al. investigated the entanglement dynamics of two particles walking on a ring, and the simulation results showed that when the coin operator includes phase parameters, the evolution period and measurement results after a finite number of steps become more complex [90]. To illustrate this further, Sun et al. considered two-particle quantum walks on a one-dimensional percolation graph, demonstrating that when in a dynamic percolation state, the joint measurement probability distribution area of the two particles is broader; while in a static percolation process, the two particles tend to cluster around their initial position [91]. Costa et al. proposed an interesting constraint for the walk process of two particles on a lattice: when the two particles collide head-on, their walking directions change [92]; this constraint is known as the HPP (Hardy, Pomeau, and de Pazzis) rule. Experimental results indicate that multi-particle quantum walks under the HPP rule are better suited for spatial search problems on lattices compared to quantum walks controlled by phase parameters and single-particle quantum walks. In another significant contribution, Carson et al. studied the entanglement dynamics of two interacting particles on regular and irregular graphs using spectral methods, and concluded that the quantum walk results of two entangled particles can effectively amplify (similar) subtle differences between subgraphs [93]. This conclusion provides a new design basis for graph isomorphism determination methods based on quantum walks. Berry et al. utilized the above conclusion to design a graph isomorphism algorithm using two-particle quantum walks [94]. Test results reveal that the quantum walk of two entangled particles can distinguish non-isomorphic strongly regular graphs, while the quantum walk of two unrelated particles can distinguish some non-isomorphic graphs with the same family parameters. Nonetheless, some studies suggest that single-particle quantum walks can also amplify subtle structural differences in graphs. As an example, Zhang et al. designed an R-convolution graph kernel based on discrete-time quantum walks [52], and experiments demonstrated that this graph kernel method can quickly and accurately distinguish strongly regular graphs that are difficult to determine using other quantum walk methods.

Other studies on multi-particle quantum walks using two particles as an example focus on designing universal quantum computing models [49]. The research approach based on multi-particle quantum walks involves defining a two-particle quantum walk model and providing the associated quantum circuit diagram, which is then simulated and verified using open-source platforms such as IBM Quantum Platform. Some studies also present proofs to validate the computational efficiency of the proposed universal design models. Additionally, some findings suggest potential application scenarios for current multi-particle quantum walks. In particular, Li et al. proposed a hash scheme based on the two-particle interactive quantum walk and demonstrated its feasibility and security [95]. The team further utilized mutual information to study the correlation between the measurement results of two interacting particles [96]. The research indicates that two-particle interactive quantum walks are beneficial for designing feasible hash schemes. These findings based on two-particle quantum walks are expected to provide new technical support for image encryption and secure communication fields.

Variants of low-dimensional quantum walks also include three-state or multi-state quantum walks [97–100]. Considering the one-dimensional three-state quantum walk, the three states refer to adding a stationary state (self-loop) at the current position, in addition to the two existing options of walking forward or backward on the line. In a one-dimensional three-state quantum walk, due to the inclusion of the stationary state, the walking distance is shorter compared to a one-dimensional quantum walk under the same conditions [99], and the average occupancy at each position on the line is lower [81]. Moreover, some studies on variants of low-dimensional quantum walks focus on entanglement entropy [101] or graphs with traps and broken links [102], aiming to explore the relationship between the measurement outcomes of quantum walks and these factors [18]. These variant studies mainly provide theoretical value, as their application scenarios remain unclear. However, research on quantum walks on graphs with broken links [102] suggests that studies on quantum walks on regular graphs are no longer sufficient for practical needs. Shifting the focus from regular graphs to irregular and complex networks has become a new research trend in the field of quantum walks.

Unless otherwise specified, quantum algorithms and computational models, including quantum walks, are generally understood as isolated quantum systems. In reality, however, quantum systems interact with their environment and cannot be completely isolated. Therefore, systems that interact with the environment are referred to as open quantum systems, and quantum walks conducted within such systems are called open quantum walks. In the study of open quantum walks, another quantum system is usually introduced as an auxiliary, forming a larger closed quantum system together with the open quantum system. For example, in 2012, Attal et al. first used quantum trajectories to simulate the evolution of open quantum walks and explored the relationship between open quantum walks on a line and one-dimensional Hadamard quantum walks in a closed quantum system [103]. Additionally, studies on variants of quantum walks include non-Hermitian or non-unitary quantum walks [104], lackadaisical (lazy) quantum walks [105], topological quantum walks [106], and aperiodic quantum walks [107]. The vast space of composite quantum systems significantly increases the computational load. As a result, the goals of the research on these variants mainly include: (1) extending from low-dimensional regular graphs to higher-dimensional graphs and providing generalized expressions; (2) implementing variant quantum walks on specific physical materials; and (3) exploring characteristics such as the limiting distribution of variant quantum walks. This book concentrates on the application of quantum walks in complex networks, and for more in-depth theoretical analysis, readers can consult [21,108,109]. Further details will not be elaborated here.

2.2 QUANTUM WALKS ON COMPLEX NETWORKS

Complex networks are characterized by irregular graph structures. Extending the application of quantum walks from regular graphs to irregular complex networks significantly alters the construction method of quantum walk operators, the motion patterns of particles on complex networks, and the interpretation of quantum walk measurement results. This section focuses on analyzing the design principles of quantum walk algorithms on complex networks and provides a general framework for quantum walk algorithms applied to complex networks.

2.2.1 Research Significance of Complex Networks

Just as particles are habitually used to describe microscopic entities in quantum mechanics, humans similarly tend to simplify and abstract complex research objects into points. For example, in swarm intelligence algorithms, the optimization process of a population is often described as the movement of points within a search space [14]. In the analysis of forces acting on objects, all information except for mass is disregarded, and the object of study is abstracted as a point mass. Building on this concept, humans can be viewed as points, and social relationships between individuals can be represented as connections between these points, thereby expressing real social relationships as a network. Similarly, proteins and the physical interlocking relationships between them can also be abstracted into a network composed of nodes and edges. Clearly, nodes and edges are the most fundamental components of a network, and in complex networks, they are referred to as nodes and links. In complex networks, nodes and links not only simplify complex entities but also more accurately represent universally interconnected phenomena.

The poet Bei Dao uses a single word, "net," to convey his profound understand about life. However, what can be described with a "net" is not limited to the intricate complexities of life alone. The transmission and feedback of information between neurons, the semantic relationships between words, the coordination and deployment of weapon systems, and the citation relationships between web pages—all these can be expressed as complex networks using nodes and links. The term "complex" is prefixed to the network because its structure exhibits diversity, repetition, self-similarity, and is often chaotic and irregular. In life, certain social events with cascading effects have highlighted the research value of complex networks. In 2003, several high-voltage lines in Ohio, USA, were burned down, causing a blackout across North America, leading to economic losses of 50 billion USD and the deaths of eight people. In September 2021, China Central Television (CCTV) aired a prime-time program called *Anchor Commentary on the News* to provide positive public guidance on international and social issues mentioned in *CCTV News*. In August 2022, the international authoritative medical journal *The New England Journal of Medicine* published a study on the discovery of a zoonotic henipavirus, bringing the advocacy of avoiding wild animal consumption back to public attention through media coverage.

The real-world cases mentioned above demonstrate that, to prevent network-wide cascading failures, it is essential to identify and provide special protection to the key nodes within the network. Likewise, effectively utilizing super-spreaders within a network can accelerate the transmission of cascading effects and amplify their impact. The advocacy of avoiding wild animal consumption illustrates that severing the links between sources of infection and other nodes within the network can effectively hinder the transmission of viruses.

The issues of node mining and link mining mentioned above are fundamental research topics in complex networks, aimed at identifying the critical few components within the network. The "butterfly effect" illustrates that any addition, deletion, or alteration of these key individuals in a complex network can lead to transformative changes in the entire system. In reality, complex networks are not absolutely static. Social networks provide a clear example: actions like account deactivation and unfollowing result in the loss of nodes and links, while the emergence of large numbers of bot accounts adds nodes to the network. Similarly, examples include neuron death in the brain's neural networks, the expansion of subway lines, and the updating of citation networks in academic papers. Apparently, the topology of complex networks evolves over time, indicating that the connections within these networks are time-sensitive. In dynamic, non-static networks, identifying the critical individuals, predicting unknown structural information, or even reconstructing the state of the network at a given moment is more aligned with practical needs. Specifically, using complex network theory to deduce protein–protein interactions, ascertain functional modules within protein networks, and locate disease-causing gene loci can save significant time and financial costs.

Following the characteristic of "describing the relationships between all universally connected entities," complex networks have become a scientific paradigm. The most intuitive example is how search engines rank search results for keywords, treating hyperlinks between web pages as links and the web pages themselves as nodes, with new and highly cited pages often ranking higher. Furthermore, complex networks can be considered as a mining tool in their own right. Their value lies in transforming unordered data into structured relational data, and leveraging complex network theory to extract other valuable information based on the graph of relational data. The practical significance of complex network research in real life is self-evident. This book will design quantum walk algorithms tailored for complex networks and introduce other innovative quantum walk algorithms used to explore structural information or characterize nodes within complex networks.

2.2.2 Overview of Quantum Walks on Complex Networks

The first study on quantum walks in complex networks can be traced back to 2006, published in the *International Journal of Quantum Information* under the title "Quantum walks on general graphs" [110]. By examining discrete-time quantum walks, this model records the irregular connectivity information between nodes in a black box and designs a shift operator based on the black box, allowing particles to move according to the link relationships between nodes in the graph. This ensures that the evolution process complies with the unitary transformation, and finally, the resource consumption of the corresponding quantum circuit is analyzed. Quantum walks on general graphs served as a starting point for the research on quantum walks in complex networks, a field that has gradually developed and continues to thrive. For quantum walks on complex networks, this section will introduce the related progress from two perspectives: discrete-time quantum walks and continuous-time quantum walks. Notably, discrete-time quantum walks on complex networks can be categorized into two types: coined quantum walks on complex networks and Szegedy quantum walks on complex networks.

2.2.2.1 Coined Quantum Walks on Complex Networks

As the name suggests, the coin operator is the core of coined quantum walks. The eigenvalues of the evolution operator, which are determined by different coin operators, lead to varying evolution outcomes and thus impact the effectiveness of solving problems. This viewpoint is strongly reflected in quantum computing research based on matrix spectral decomposition methods [3,22,111]. In coined quantum walks on complex networks, the Grover operator is the most widely used coin operator. Schofield et al., based on the original definition of the Grover operator, incorporated node degree values and adjacency relationships into the coin operator [112]. This quantum walk approach can solve the NP-complete graph isomorphism problem with a time complexity of $O(N^9)$, where N is the number of nodes in the network. Wang et al. combined the standard basis of nodes with the definition of the Grover operator to define quantum walks driven by the Grover coin operator, applying it to node similarity assessment [113] and role embedding [114]. This coin operator was also adopted in [115]. Mukai et al. defined both Grover and Fourier coin operators on complex networks, projecting the real and imaginary parts of the eigenvalues of different evolution operators as scatter points onto a unit complex plane, presenting them in the form of a unit circle [111]. Moreover, the study in [111] hypothesized that when the scatter point distribution pertaining to the eigenvalues is more uniform, the coin operator is more effective in detecting community modules within complex networks. This hypothesis was validated on the well-known open-source Karate Club social network and the US airline network. Chawla et al. constructed a parameterized coin operator based on the out-degree and in-degree information of nodes, ranking the importance of nodes within complex networks using one-dimensional discrete quantum walks [116].

2.2.2.2 Szegedy Quantum Walks on Complex Networks

Another significant discrete-time quantum walk model on complex networks is the Szegedy quantum walk, which incorporates node weight information from the network into the initial probability amplitudes and ensures that the evolution satisfies the unitary transformation. This model is also a well-known quantum algorithm based on Markov chains with accelerated search properties [37]. In contrast to coined quantum walks, Szegedy quantum walks are coinless quantum walks. Wong further revealed the equivalence between Szegedy quantum walks and coined quantum walks in querying marked points [117], highlighting exceptions for the former in the case of a one-dimensional line [118], thereby providing a clearer understanding of the applications and limitations of Szegedy quantum walks. The emergence of Szegedy quantum walks offers new solutions for network analysis problems. In 2013, Paparo et al. extended Google's famous PageRank algorithm into a quantum version based on this quantum walk [119,120]. Experimental results indicated that the quantum PageRank algorithm not only effectively ranks the importance of nodes within the network but also clearly identifies the hierarchical structure of the network. Loke et al. compared the classical and quantum PageRank algorithms across different types of random networks and validated that the quantum algorithm effectively distinguishes marginal nodes within networks [121]. Furthermore, Bai designed a high-order weighted paper impact evaluation method based on Szegedy quantum walks [122], which effectively identifies key nodes in citation networks. Wang proposed a dual-function search and ranking integration algorithm, embedding the evolution operator of the quantum PageRank algorithm with marked node information to create an evolution operator with search functionality [33]. Experimental results showed that this algorithm can efficiently rank key nodes within the network while rapidly searching for marked nodes.

2.2.2.3 Continuous-Time Quantum Walks on Complex Networks

Initially, the integration of continuous-time quantum walks with complex networks was full of uncertainties and challenges. In 2008, Xu et al. grounded in ER (Erdös-Rényi) random networks, concluded that in the probability distribution of continuous-time quantum walks, the measurement probability at the particle's initial position was disproportionately high [123]. Faccin et al. compared the probability distribution characteristics of continuous-time quantum walks on complex networks with classical continuous-time random walks [124]. This study reached an interesting conclusion regarding network structure mining: if the Hamiltonian of a continuous-time quantum walk is the graph's normalized Laplacian matrix, it is more effective in identifying the key nodes of the network. Similar exploratory studies can be found in [125] and [126], where the researchers studied network transmission efficiency using continuous-time quantum walks and evaluated the differences with classical random walks. In 2014, Faccin et al. further developed a formula for node closeness based on continuous-time quantum walks to detect community structures in networks [127]. Experimental results demonstrated that the measurement outcomes of quantum walks could reflect the structural features of networks.

As a general computational model for quantum computing, the application of continuous-time quantum walks on complex networks often evokes associations with spatial search problems. Chakraborty et al. proposed an interpolated Markov chain based on continuous-time quantum walks that could search for a single marked point on any graph, with a search efficiency equivalent to the square root of the time required by classical search algorithms [128]. For scale-free networks, Osada et al. discovered that when using continuous-time quantum walks to search for marked points, the search efficiency depends on whether the marked node is a central node [129]. Li et al. improved the efficiency of searching for multiple marked points in strongly regular graphs by adding self-loops [130], showing that when the jump rate of the continuous-time quantum walk is set to $1/k$, where k is the degree of the strongly regular graph, the method can search for multiple marked points within $O(\sqrt{N})$. Chakraborty et al. demonstrated that for an ER network with connection probability p, when $p \geq \log^{3/2}(N)/N$, continuous-time quantum walks provide the optimal method for searching marked points, where N is the number of nodes [75]. These findings provide a solid theoretical foundation for the application of continuous-time quantum walks in the field of complex network structure mining.

Another notable achievement is the quantum PageRank algorithm based on continuous-time quantum walks [131]. Although both this and the quantum PageRank algorithm derived from Szegedy quantum walks can identify key nodes in complex networks, the former belongs to the category of open-system quantum walks, and its simulation on existing computers incurs significant computational costs [121].

2.2.2.4 Review of Quantum Walks on Complex Networks

As described above, the research on the application of quantum walks in complex networks is still in its nascent stage and has certain limitations. This section provides an analysis of these shortcomings.

In pursuit of minimizing runtime on quantum devices, researchers have lowered the precision requirements for solving network structure mining problems using quantum walks. Constrained by conditions such as unitary transformations, existing algorithms can only solve problems using simplistic quantum walk models, merely evaluating whether the measurement outcomes of a given quantum walk model are useful for the target task [132]. For example, in the case of continuous-time quantum walks, this application directly incorporates the network adjacency matrix into the solution of the Schrödinger equation to obtain the evolution operator. The importance of different nodes in the network is then assessed by comparing measurement outcomes. In practice, however, the measurement results of quantum walks are influenced by multiple factors, including the initial position of the walk, the walk steps, and whether self-loops are present, among other conditions. These factors have not been fully exploited in existing studies, leading to unsatisfactory computational results for quantum walk algorithms when applied to complex network problems.

Additionally, designing quantum walk algorithms on complex networks for quantum computers faces numerous constraints, while the requirement for solving practical problems is often urgent. Retaining the theoretical advantages of quantum computing while appropriately discarding certain constraints of quantum algorithms could provide new approaches for designing algorithms to mine structural information in complex networks. Specifically, quantum walk algorithms take the adjacency matrix of a network as input, effectively obtaining global information about the network, which typically offers higher precision than local metrics. Integrating local information appropriately into this global information can significantly enhance the accuracy of complex network structure mining tasks. The quantum measurement process is similar to scoring nodes or links in a given problem, and known information can be incorporated during this process to adjust the scores of nodes or links, thereby improving solution accuracy. Compared to solving application problems in complex networks, the development of quantum computers will likely take a much longer time. Consequently, it may be feasible to create network structure mining algorithms by retaining only the advantageous aspects of quantum computing, providing new perspectives for applying quantum walks in complex networks. Building on these ideas, this book will design several quantum walk algorithms aimed at mining meaningful structural information in complex networks, and it will introduce representative and innovative applications of quantum walks in complex networks.

2.2.3 Design of Quantum Walk Algorithms on Complex Networks

Based on the experimental analysis in Section 2.1 and the review of quantum walk algorithms on complex networks in Section 2.2.2, this section further clarifies how to design quantum walk algorithms for complex networks. This part focuses on analyzing the walk steps, initial probability amplitude, evolution operator, and the new challenges that arise when extending quantum walks to complex networks, providing insightful and thought-provoking considerations.

2.2.3.1 Setting the Walk Step

First, it is obvious that the walk steps of quantum walks on complex networks cannot approach infinity and then be averaged [132]. On one hand, the enormous computational resource consumption resulting from infinite walk steps would render quantum walk algorithms on complex networks infeasible. On the other hand, the periodic patterns in the measurement outcomes of quantum walks on complex networks become extremely complex or even immeasurable with infinite walk steps. Taking discrete-time quantum walks on complex networks as an example, each step of the particle's walk is measurable, equivalent to the particle executing a breadth-first search from the starting node. Apparently, the walk steps should not approach infinity and then be averaged. In contrast, continuous-time quantum walks on complex networks have their own particularities. Since it is not possible to observe walk outcomes at intervals of "one step," the periodic changes in the probability amplitude at a node become the primary basis for measurement. The experimental results in Figure 2.6 indicate that the walk distance aligns with the range of probability amplitude propagation from the initial point, which is similar to information diffusion—information can only diffuse within an effective range. Hence, the time parameter for continuous-time quantum walks on complex networks (corresponding to the walk steps in discrete-time quantum walks) should also be a relatively small value. Furthermore, based on the experimental results in Figure 2.2, different walk steps result in diverse eigenvalues of the evolution operator. Eigenvalues are directly related to the accuracy of identifying key nodes (or groups) within the network system [133]. Therefore, the walk steps should not only be limited but also be set based on specific criteria. This book will utilize the above research approach to introduce an information propagation model based on continuous-time quantum walks in Chapter 3 and present a three-degree decay Grover walk algorithm for identifying key nodes in networks.

2.2.3.2 Setting the Probability Amplitude

Considering the experimental results discussed in Section 2.1, it is not difficult to reach the following conclusion: the preset initial probability amplitude can determine the particle's walk path. In complex networks, the reference value of this conclusion for quantum walk algorithm design should be considered separately from both discrete and continuous perspectives. Comparing the dimension of the Hilbert space with the number of network nodes reveals that the evolution of discrete-time quantum walks on complex networks is an expansive transformation, whereas the evolution operator of continuous-time quantum walks maintains the same matrix size, suggesting its evolution is non-expansive. Obviously, when using discrete-time quantum walks to mine network structure information, setting different initial probability amplitudes for each node to define their walk paths and then measuring the probability of the particle staying at each node would consume a large amount of computation. In contrast, continuous-time quantum walk algorithms are more suitable for assigning independent initial probability amplitudes to each node, as the computational load is much lower than that of discrete-time quantum walks. In other words, discrete-time quantum walk algorithms should maintain the conventional setting of equal superposition of quantum states.

2.2.3.3 Construction of the Evolution Operator

Informed by the experiments in Section 2.1, it is evident that certain characteristics reflected by different coin operators in low-dimensional discrete-time quantum walks become irrelevant in complex networks.

For instance, in complex networks, each step of a discrete-time quantum walk is equivalent to performing a BFS (breadth-first search), and this property does not change regardless of the coin operator used. Thus, the difference in walk distances resulting from different coins loses its significance. Generally, the role of the coin operator in complex networks remains to determine the next step's direction, but it should be clearly distinguished as either biased or unbiased, and the appropriate coin should be chosen based on the specific research problem. For example, in the task of identifying key nodes, the degree of a node is a straightforward local evaluation metric. From this perspective, a biased Grover matrix that incorporates node degree information may be more suitable as a coin operator. Similar applications include the use of biased Fourier coins in studies of community structure detection in networks [111]. For the task of critical link mining discussed in this book, an unbiased Hadamard operator is more appropriate as the coin operator for the walk. This is because evaluating the importance of a link primarily depends on node similarity information, and a biased coin may not effectively reflect this feature.

Furthermore, the evolution operator for discrete-time quantum walks includes a shift operator to facilitate particle movement between different nodes. Complex networks are irregular graph structures, and the shift operator in quantum walks on complex networks must perform either a flip or swap operation to satisfy the conditions of unitarity, meaning that the movement process between nodes j and k must be reversible (with nodes j and k being neighbors). Therefore, unless otherwise specified, the complex networks referred to in this book are undirected complex networks.

2.2.3.4 Analysis of Derived Issues

The movement of particles between different nodes during quantum walks can be viewed as the traversal of the graph structure by the particle. Using a two-step walk as an example, the traversal results of classical random walks and quantum walks can be seen in Figure 2.7. For the same two-step walk, the number of nodes visited by the quantum walk is 7. Moreover, one can imagine that when the walk steps exceed the network's diameter, the quantum walk will exhibit a considerable amount of revisiting behavior. From low-dimensional quantum walks to quantum walks on complex networks, one of the primary issues faced by quantum walk algorithms when applied to network structure information mining is the negative impact caused by repeated visits. Following the convention used in classical random walks, this negative impact is understood as a traceback. Given the

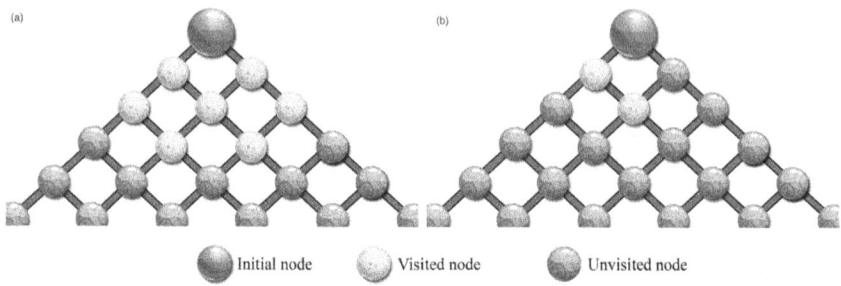

FIGURE 2.7 Traversal results of classical random walks and quantum walks after two steps (a and b).

acceleration characteristics of quantum walks in spatial search, quantum walks traverse network nodes faster than classical random walks, making the negative effects of traceback even more pronounced. This leads to fluctuating measurement outcomes for the particle as the walk steps changes, deviating from the evaluation criteria that are favorable for network structure information mining.

The occurrence of a traceback is closely related to the interference effect of quantum superposition and is embedded in the calculations through concepts like probability amplitude and standard basis in the form of matrix multiplication. Therefore, traceback cannot be entirely eliminated in quantum algorithms; it can only be mitigated through interventions to reduce its negative impact. This book discusses four methods to minimize the adverse effects of traceback: specifying the walk path (see Section 3.3.3), setting a meaningful and extremely short walk steps (see Section 3.2.3), adding self-loops to nodes (see Sections 3.2.3 and 4.3.2), and using measurement-free quantum walks (see Section 5.2.1). Apart from tasks focused on mining structural information in complex networks, overcoming the negative effects of traceback in quantum walks on complex networks is another central theme throughout the quantum algorithm designs presented in this book.

2.3 GENERAL FRAMEWORK OF QUANTUM WALK ALGORITHMS IN THIS BOOK

To facilitate the description of quantum walk algorithms on complex networks in subsequent chapters, this section designs and introduces a general framework for quantum walk algorithms on complex networks, based on the relationship between the fundamental assumptions of quantum mechanics and non-quantum algorithm design. Section 1.1.4 introduces the four basic assumptions of quantum mechanics. If quantum walk algorithms on complex networks are defined according to the content and order of these four basic assumptions, there will be significant overlap and intersection in the definitions and descriptions. For instance, defining the state space begins with establishing the Hilbert space that constrains the particle's motion. The Hilbert space, in turn, is determined by the selectable directions of particle movement on the network. Meanwhile, the construction of the evolution operator relies, on one hand, on the dimensional information provided by the Hilbert space, and on the other hand, requires that the particle's movement between nodes on the graph satisfies the constraints of unitary transformations. The composite system indicates that the Hilbert space on the complex network is composed of network nodes and the neighborhood information of those nodes, and its structure depends on the product of the shift operator and the coin operator. Clearly, the related definitions are intertwined and repetitive in their content, so a unique design framework for quantum walk algorithms on complex networks needs to be provided. Figure 2.8 explains the content and design basis of this framework.

First, the primary requirement for quantum walk algorithms on complex networks is to define the Hilbert space on the complex network. Once this space is identified, it implies the following: (1) The number of walkable directions at each node in the network is determined based on the

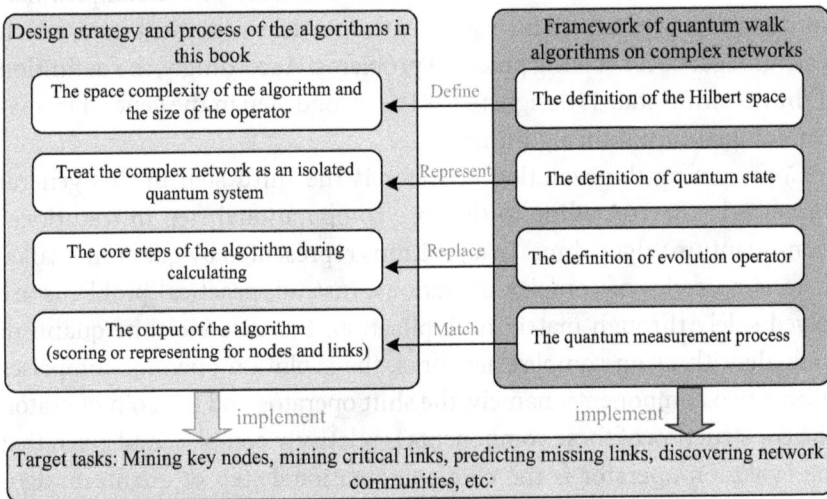

FIGURE 2.8 General framework of quantum walk algorithms on complex networks.

number of nodes or their degree values. (2) The length of the quantum state must be equal to the dimension of the Hilbert space. (3) The evolution operator, as the core of the quantum walk algorithm, should have row and column dimensions identical to those of the Hilbert space to satisfy the requirements for matrix multiplication. (4) The space complexity of the algorithm is thus ascertained, as the simulation of quantum walks on current computers must be expressed through matrix and vector calculations. In alignment with the basic postulates of quantum mechanics discussed in Section 1.1.4, defining the Hilbert space as the first step in the general framework of quantum walk algorithms provides the necessary elements for state space, unitary evolution, and composite systems.

Based on the established Hilbert space of the complex network, the second step of this framework is to define the initial quantum state, including the standard basis and probability amplitude of the complex network nodes. As an integral part of the quantum state, the probability amplitude can be designed according to the requirements of the research problem. For instance, under the condition that the sum of the squares of the probability amplitudes equals 1, a probability amplitude distribution scheme tailored to practical requirements can be designed. In this case, the initial probability amplitude in the quantum state describes the state of the particle residing at different nodes at the initial moment, directly affecting the quantum measurement outcome. By specifying the quantum state in this manner, it not only represents the overall structure of the complex network under study but also incorporates the particle movement state, which is advantageous for solving practical problems. Accordingly, the definition of the quantum state is designated as the second step in the general framework of quantum walk algorithms.

Constructing the evolution operator is the third step in this general framework, corresponding to the core computational step in traditional (non-quantum) algorithms. In algorithms represented by quantum walks, there is no series of complex operations; instead, practical problems are solved solely through matrix multiplication. For discrete-time quantum walk algorithms on complex networks, the evolution operator comprises at least two components, namely, the shift operator and the coin operator. The construction of these components is relatively complex, and given that the evolution operator is the core computational step of quantum algorithms, it is designated as the third step in the algorithm design framework.

Finally, the measurement process corresponds to the output of the algorithm, making it the fourth step in the framework. As previously emphasized, matrices are the core expression and computational form of quantum algorithms. As a result, to score each node in the complex network based on the specific requirements of the task, nodes must be mapped either in groups to the matrix components (in cases where some matrices represent expansive transformations) or in a one-to-one manner. This scoring of nodes, links, or subgraph structures is employed to accomplish various structural mining tasks on complex networks.

2.4 SUMMARY

This chapter presents the foundational theory of quantum walk on regular graphs and provides a comprehensive overview of the research on quantum walk in complex networks as well as those on regular graphs. By exploring the experimental phenomena of quantum walk on regular graphs, the chapter analyzes, from multiple perspectives, the design concepts of quantum walk algorithms in complex networks, ultimately offering a general framework for describing and defining quantum walk algorithms in these networks. The subsequent chapters of this book will utilize this framework to introduce the discrete-time quantum walk algorithms on complex networks.

II

Applications

CHAPTER 3

Applications of Quantum Walks in Node Discovery

3.1 NODE DISCOVERY: DEFINITIONS AND EVALUATION METRICS

Nodes are one of the basic components of complex networks. Suppose a node in the center of a complex network is removed, the network may disintegrate into several independent subnetworks. If a node at the edge of the complex network (a dangling node) is deleted, the impact on the network is minimal. This shows that the importance of nodes in a network varies, and the addition or removal of key nodes can bring about significant changes, possibly even catastrophic, to the entire network system. Therefore, identifying key nodes in complex networks has become a core task in applications such as viral marketing, public opinion diffusion, and power grid protection [134–136]. In complex networks, super-spreaders, high-influence nodes, seed nodes, and central nodes are all equivalent to the concept of key nodes, and node ranking is the same task as key node mining. Specific evaluation methods are necessary to quantify whether a node is a critical individual. Common evaluation methods include the susceptible-infected-recovered (SIR) model [137], robustness index [137], Kendall coefficient [138], and influence maximization [139].

The evaluation method of the SIR model assesses the importance of nodes in a network by simulating the spread of infectious diseases [137]. In this model, a node's state can only be one of three: susceptible, infected,

DOI: 10.1201/9781003683902-5

or recovered. Assuming the node u to be evaluated is an infected node, it spreads the disease cascade-like to its neighboring nodes under a preset infection probability. There is a possibility that infected nodes may recover during this process. The final number of infected nodes in the network serves as an important score for node u. This process is stochastic and is usually repeated 10^3 times, with the average taken as the final score for node u. The higher the average, the greater the node's influence, indicating that it is more crucial in the network.

The robustness index calculates the degree of disintegration of the entire network after the removal of the evaluated node. If removing node u results in the network being fragmented into several independent subgraphs, node u is particularly significant within the network; if the network structure remains intact, the importance of node u in the overall network system is relatively low. Given a complex network $G=(V,E)$, where V and E represent the set of nodes and links of the network, respectively, and $E \subseteq V \times V$, if the number of nodes in network G is N and the number of links is M, that is, $|V| = N$ and $|E| = M$. Assuming that after the deletion of the ξth node, the degree of disintegration of the network is expressed by the ratio $\sigma(\xi/N)$ of the number of nodes in the largest connected component to the total number of nodes in the network. For the entire network system, after removing each node from the network one by one, the robustness index can be defined as

$$R = \frac{1}{N} \sum_{\xi=1}^{N} \sigma\left(\frac{\xi}{N}\right), \tag{3.1}$$

where $1/N$ is used for normalization of the measurement results, hence $R \in (0,1)$. The smaller the value of $\sigma(\xi/N)$, the smaller the scale of the connected component obtained by the network disintegration after the removal of the node, and the node with the smaller value of R is more important in the complex network.

The Kendall coefficient is an indirect evaluation metric that uses existing algorithms to evaluate the ranking ability of a given algorithm on key nodes. It judges the performance of an algorithm in accessing key nodes by the correlation between the measurement values of key nodes via different algorithms. Simply put, when using the Kendall coefficient to estimate the accuracy of an algorithm in assessing the importance of nodes, the measurement results of two algorithms for nodes should be regarded as two distributions, denoted as X and Y, where $X = \{x_i | i = 1, 2, ..., N\}$, $Y = \{y_i | i = 1, 2, ..., N\}$. If a pair of measurement values (x_j, y_j) and (x_k, y_k) are positively correlated, then the counter N_c of the positive correlation statistic is incremented by 1; if they are negatively correlated, then the negative correlation statistic N_d is incremented by 1. Thus, the Kendall coefficient τ can be defined as

$$\tau = \frac{N_c - N_d}{\frac{1}{2}N(N-1)},$$ (3.2)

where the calculation methods for N_c and N_d are

$$\begin{cases} N_c = N_c + 1, \left[(x_j > x_k) \cap (y_j > y_k)\right] \cup \left[(x_j < x_k) \cap (y_j < y_k)\right] = T \\ N_d = N_d + 1, \left[(x_j < x_k) \cap (y_j > y_k)\right] \cup \left[(x_j > x_k) \cap (y_j < y_k)\right] = T \end{cases}.$$

(3.3)

Influence maximization is a classic application for evaluating the influence of network nodes and is classified as an NP problem. The research objective is to identify highly influential seed nodes to maximize the spread of information coverage in the network [139]. Let the influence propagation function (also known as the extent function) be denoted as $\delta(\cdot)$, with the input seed set S, where the size of this seed set is less than the budget l, that is, $|S| \leq l$, with $l \ll N$ and $S \subseteq V$. The objective function for influence maximization [134] is defined as

$$S^* = \underset{|S| \leq l; S \subseteq V}{\arg\max} \delta(S).$$ (3.4)

As the size of the seed set increases, the propagation results of the influence also grow, and this growth follows the principle of marginal effects, meaning that the incremental results of propagation decrease as the size of the seed set expands. Consequently, the objective function in Eq. (3.4) possesses properties of monotonicity, non-negativity, and submodularity [136]. These three properties collectively determine that the influence maximization application examines the performance of the algorithm in mining key nodes based on the maximization of the propagation results from seed nodes.

The four methods mentioned above serve as the evaluation basis for the quantum walk algorithm in this chapter for the task of mining key nodes in complex networks.

3.2 DISCRETE-TIME QUANTUM WALKS IN NODE DISCOVERY

This section introduces three discrete-time quantum walk (DTQW) algorithms for mining key nodes in networks, all of which are described following the general framework presented in Section 2.3.

3.2.1 Quantum PageRank Algorithm

Currently, the quantum version of Google's PageRank algorithm includes both discrete-time [119] and continuous-time versions [131]. This section presents the former—the quantum PageRank algorithm [121] based on Szegedy's quantum walk [37], referred to as the QPageRank algorithm. In the process of quantizing the classical PageRank algorithm by using Szegedy's quantum walk and the probability transition matrix, the QPageRank not only addresses the challenge of ensuring that the evolution meets the unitary condition due to the introduction of the probability transition matrix as an initial parameter in quantum walks, but also can achieve accurate ranking of central nodes in complex networks.

Following the general framework for DTQW algorithms in Section 2.3, the Hilbert space $\mathcal{H}$ of the QPageRank algorithm is established. Since Szegedy's quantum walk is defined based on Markov chains, it transforms the link relationships of nodes in the network graph into a bipartite graph [37], where each part of the bipartite graph consists of the node set V, and $|V| = N$, therefore, the Hilbert space $\mathcal{H}$ for the QPageRank algorithm is constructed as

$$\mathcal{H} = \mathcal{H}^N \otimes \mathcal{H}^N. \tag{3.5}$$

According to the operational characteristics of the tensor product in Eqs. (3.5) and (1.8), for the complex network G, the dimensionality of the Hilbert space of the QPageRank algorithm is N^2. Using the standard basis $|j\rangle$ to express any node $j \in V$ in the network, and $|j,k\rangle$ to represent the link relationship between node j and node k, it is possible to construct the initial state $|\psi_j\rangle$ of any node j in the complex network. This process uses the probability transition matrix to set all initial probability amplitudes:

$$|\psi_j\rangle = |j\rangle \otimes \sum_{k=1}^{N} \sqrt{P_{kj}^{G}} |k\rangle$$

$$= \sum_{k=1}^{N} \sqrt{P_{kj}^{G}} |j,k\rangle,$$

(3.6)

where P_{kj}^{G} signifies the element in the kth row and jth column of the probability transition matrix. By superimposing the initial states of all nodes, the initial quantum state of the QPageRank algorithm for the network system G can be obtained, calculated as

$$|\psi_0\rangle = \frac{1}{\sqrt{N}} \sum_{j=1}^{N} |\psi_j\rangle.$$

(3.7)

Subsequently, the evolution operator of the QPageRank algorithm is defined, which is composed of a projection operator and a shift operator. Referring to the description of the reflection operator in the Grover search algorithm in Section 1.2.1, the projection operator $\hat{\Pi}$ is described as

$$\hat{\Pi} = \sum_{j=1}^{N} |\psi_j\rangle\langle\psi_j|.$$

(3.8)

The shift operator in the QPageRank algorithm represents jumping from the current node to a node in its neighborhood. This process is reversible, so the shift operator is defined as

$$\hat{S} = \sum_{j,k=1}^{N} |j,k\rangle\langle k,j|.$$

(3.9)

Based on the projection operator in Eq. (3.8) and the shift operator in Eq. (3.9), the evolution operator for the QPageRank algorithm is expressed as

$$\hat{U} = \hat{S}(2\hat{\Pi} - \hat{I}),$$

(3.10)

where $\hat{I}$ denotes the identity matrix. The QPageRank algorithm employs a two-step evolution, calculated as

$$\hat{U}^2 = (2\hat{S}\hat{\Pi}\hat{S} - \hat{I})(2\hat{\Pi} - \hat{I}).$$

(3.11)

The purpose of using two-step evolution is to ensure that the evolution process of the QPageRank algorithm satisfies the conditions for unitary transformation [117,121]. Finally, after t steps of walking, the scoring of webpage j or network node j is realized through the following measurement process:

$$\text{Pro}(j,t) = \left\langle \psi_0 \left| \hat{U}^{\dagger 2t} \right| j \right\rangle_2 \left\langle j \left| \hat{U}^{2t} \right| \psi_0 \right\rangle, \tag{3.12}$$

where $\hat{U}^{\dagger 2t}$ is the conjugate transpose of $\hat{U}^{2t}$.

The QPageRank algorithm provides important references for the application of quantum walks in complex networks. For example, Wang proposed an integrated algorithm based on the QPageRank algorithm that can both rank key nodes and search for marked points [33]; Bai used this algorithm to mine key nodes in citation networks [122]. The QPageRank algorithm not only extends the PageRank algorithm to run on quantum computers but also solves the problem of distinguishing the influence of nodes with lower rankings in the PageRank algorithm. Consequently, it is considered one of the more successful algorithms in quantum computing.

3.2.2 Parameterized Coin Quantum Walk Algorithm

The coined quantum walk with parameters (CQWP) algorithm, defined by Chawla et al. [116], is based on one-dimensional DTQWs. This algorithm incorporates the outgoing degree and incoming degree information of complex network nodes into the coin operator during evolution. The results show that this algorithm is suitable for identifying key nodes in directed acyclic complex networks. Following the general framework in Section 2.3, we first introduce the Hilbert space for this algorithm. According to Section 2.1.1, the dimension of the Hilbert space for a one-dimensional quantum walk is related to the number of directions a particle can choose to walk (which is 2 on a straight line) and the total number of network nodes, thus

$$\mathcal{H} = \mathcal{H}^2 \otimes \mathcal{H}^N. \tag{3.13}$$

According to Eq. (3.13), the row and column numbers of the initial state vector and the evolution operator are both $2N$. Sequentially, the initial state vector $|\psi(0)\rangle$ of the CQWP algorithm is defined as

$$|\psi(0)\rangle = \left(\frac{|0\rangle + |1\rangle}{\sqrt{2}} \right) \otimes \sum_{j=1}^{N} \frac{1}{\sqrt{N}} |j\rangle, \tag{3.14}$$

where $|0\rangle$ and $|1\rangle$ illustrate the two walking directions on a one-dimensional line. Further, based on the product of the coin operator and the shift operator, the evolution operator of the algorithm is provided. Referencing the quantum walk on a one-dimensional line, the shift operator is defined as

$$
S_{\pm} = \begin{cases} \sum_{j=1}^{N} |0\rangle\langle 0| \otimes |j \pm 1\rangle\langle j| + |1\rangle\langle 1| \otimes |j\rangle\langle j| \\[2em] \sum_{j=1}^{N} |0\rangle\langle 0| \otimes |j\rangle\langle j| + |1\rangle\langle 1| \otimes |j \pm 1\rangle\langle j| \end{cases} . \tag{3.15}
$$

The coin operator in the CQWP algorithm contains the degree information of nodes. Combining the construction form of the SU(2) unitary matrix (see Eq. (1.16)), the coin operator is formulated as

$$
C = \sum_{j=1}^{N} \begin{pmatrix} \sqrt{\dfrac{1}{\alpha_j + 1}} & \sqrt{\dfrac{\alpha_j}{\alpha_j + 1}} \\[1.5em] \sqrt{\dfrac{\alpha_j}{\alpha_j + 1}} & -\sqrt{\dfrac{1}{\alpha_j + 1}} \end{pmatrix} \otimes |j\rangle\langle j|, \tag{3.16}
$$

where $|j\rangle\langle j|$ is used to match node j with its corresponding degree information; α_j is related to the degree value and weight of the node, defined as

$$
\alpha_j = \frac{|N_{\text{in}}(j)|}{|N_{\text{in}}(j)| + |N_{\text{out}}(j)|}, \tag{3.17}
$$

where $N_{\text{in}}(j)$ and $N_{\text{out}}(j)$ depict the incoming and outgoing neighbor sets of node j, respectively. Thus, the evolution operator is the product of the shift operator and the coin operator, namely $S_{\pm} \cdot C$. The CQWP algorithm stipulates that when calculating the importance of network nodes using the above definitions, applying the evolution operator 50 times and then averaging the results will yield a convergent ranking [116].

Since both the QPageRank algorithm (Section 3.2.1) and the CQWP algorithm in this section are benchmarked against the classical PageRank, this section evaluates their performance in ranking key nodes via the same network data. Using the network in Figure 3.1a as experimental data,

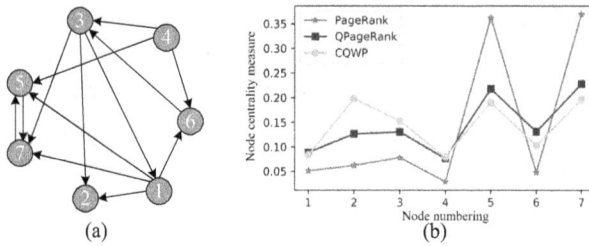

FIGURE 3.1   Network data and experimental results for QPageRank, CQWP, and PageRank. (a) Network data; (b) node centrality measurement.

TABLE 3.1   Comparison of Results Calculated by QPageRank, CQWP, and PageRank Algorithms

Nodes	QPageRank	CQWP	PageRank
1	0.0890 (0.00217)	0.0838 (0.00185)	0.0510
2	0.1265 (0.00503)	0.1984 (0.00658)	0.0619
3	0.1306 (0.00403)	0.1529 (0.00444)	0.0779
4	0.0076 (0.00146)	0.0796 (0.00251)	0.0289
5	0.2177 (0.01110)	0.1903 (0.00723)	0.3624
6	0.1313 (0.00494)	0.1034 (0.00323)	0.0480
7	0.2282 (0.01054)	0.1967 (0.00791)	0.3699

Figure 3.1b presents the ranking results of node measurement values for the three algorithms. Accordingly, Table 3.1 shows the measurement values of different nodes for the QPageRank algorithm and the CQWP algorithm, where the values in parentheses are the standard deviations of the importance measurement values of the same nodes by the current algorithm and the PageRank algorithm. The experimental results demonstrate that the ranking results of the QPageRank algorithm and the CQWP algorithm are the same, but there are slight differences between their ranking results and those of the PageRank algorithm. This indicates that, with the ranking results of the PageRank algorithm as a reference, both algorithms provide effective and novel quantum methods for key node ranking applications. However, it also suggests that the ranking results of DTQWs for key nodes are not completely dependent on network topology.

3.2.3 Three-Degree Decay Grover Walk Algorithm

This section discusses a quantum walk algorithm driven by the Grover coin operator [140]. Inspired by the three-degree influence principle, the algorithm innovatively controls the quantum walk step within 3 steps

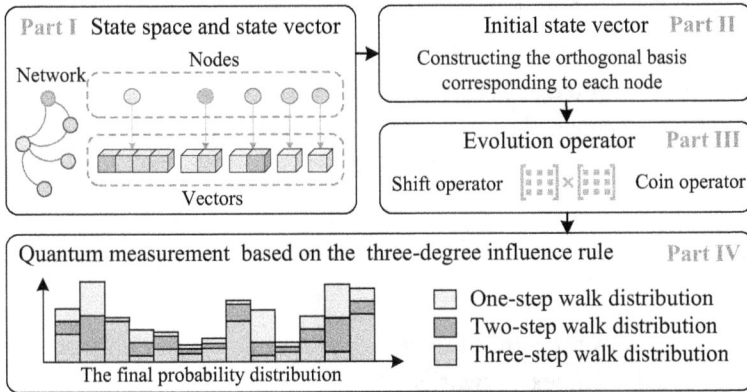

FIGURE 3.2   Framework of the three-degree decay Grover walk algorithm.

and accumulates the measurement results of 1-step, 2-step, and 3-step walks with a decay coefficient. This algorithm is hereafter referred to as the three-degree decay Grover walk algorithm. Based on the experimental results of the Kendall coefficient and robustness index based on the SIR model, the proposed three-degree decay Grover walk algorithm can reduce the negative effects of traceback and accurately mine key nodes in complex networks. According to the general framework in Section 2.3, Figure 3.2 provides a framework diagram of the three-degree decay Grover walk algorithm.

3.2.3.1 Design of the Three-Degree Decay Grover Walk Algorithm

First, for each node in the complex network G, a self-loop is added to obtain the network $G' = (V, E')$. Since $|V| = N$ and $|E| = M$, it follows that $|E'| = M + N$. The role of the self-loop is similar to the additional third state introduced in the three-state quantum walk [97]. It can increase the probability of the particle remaining at the initial node in the measurement results, thereby indirectly reducing the negative impact of traceback. In line with the order of definitions in the general framework of Section 2.3, this section considers using the direct sum operation to accumulate the space associated with each node, defined as

$$\mathcal{H} = \mathcal{H}_1 \oplus \mathcal{H}_2 \oplus \cdots \oplus \mathcal{H}_N, \tag{3.18}$$

where $\mathcal{H}$ consists of the Hilbert space $\mathcal{H}_j$ corresponding to node j, $\forall j \in V$. The dimension of the Hilbert space $\mathcal{H}_j$ is the number of the nearest neighbors of node j, denoted as $|N(j)|$, then the dimension $D_{\mathcal{H}}$ of the total Hilbert space $\mathcal{H}$ is characterized as

$$D_{\mathcal{H}} = \sum_{j=1}^{N} |N(j)|, \tag{3.19}$$

where $D_{\mathcal{H}}$ denotes the sum of degrees of all nodes in the network, hence for undirected complex networks, $D_{\mathcal{H}} = 2M + N$. The second part of the three-degree decay Grover walk algorithm pertains to the state vector. The state vector at the initial moment is defined as follows:

$$|\psi(0)\rangle = \sum_{j=1}^{N} \sum_{k=1}^{N(j)} \alpha_{j,k}(0) |j,k\rangle$$

$$= \frac{1}{\sqrt{N}} \sum_{j=1}^{N} \sum_{k=1}^{N(j)} \frac{1}{\sqrt{|N(j)|}} |j,k\rangle. \tag{3.20}$$

Here, $\alpha_{j,k}(0)$ signifies the probability amplitude of the particle transitioning from node k to node j at the initial moment, where $\alpha_{j,k} \in [0, 1]$. The transition of the particle from node j to node k is represented by a standard basis element $|j,k\rangle$. The sum of the squared modulus of the probability amplitudes on a complex network remains 1, which meets the following requirement:

$$\sum_{j=1}^{N} \sum_{k=1}^{N(j)} |\alpha_{j,k}(t)|^2 = 1. \tag{3.21}$$

The Hilbert space for the three-degree decay Grover walk algorithm is composed of the direct sum of the neighborhood of each node; thus, when defining the coin operator of the algorithm, each node has an independent coin operator, referred to as the local coin operator. Following the form of the Grover operator [2], the local coin operator is described as

$$C_j \begin{pmatrix} |j,k_1\rangle \\ |j,k_2\rangle \\ \vdots \\ |j,k_{d_j}\rangle \end{pmatrix} = \frac{1}{d_j} \begin{pmatrix} 2-d_j & 2 & \cdots & 2 \\ 2 & 2-d_j & \cdots & 2 \\ \vdots & \vdots & \ddots & \vdots \\ 2 & 2 & \cdots & 2-d_j \end{pmatrix} \begin{pmatrix} |j,k_1\rangle \\ |j,k_2\rangle \\ \vdots \\ |j,k_{d_j}\rangle \end{pmatrix},$$

$$(3.22)$$

where $d_j = |N(j)|$. Subsequently, utilizing the form of Eq. (3.18), the global coin operator is constructed based on the local coin operator, determined as

$$C = C_1 \oplus C_2 \oplus \cdots \oplus C_N. \qquad (3.23)$$

Since complex networks are irregular graphs, the movement of particles on complex networks still needs to satisfy the characteristic of reversibility. Therefore, the process of the particle transitioning from node i to node k through the action of the shift operator must be mutually exchangeable, thereby meeting the evolution condition of unitary transformation. Thus, the action of the shift operator is defined as

$$S|j,k\rangle = |k,j\rangle. \qquad (3.24)$$

Finally, the quantum measurement process of the three-degree decay Grover walk algorithm is designed based on the three-degree influence principle, as illustrated in Part 4 of Figure 3.2. The three degrees of influence rule [141] is a well-known conclusion in social statistics that reflects how effective influence propagation occurs only within three degrees of separation. For example, considering node u, Figure 3.3 highlights the specific meanings of the first, second, and third degrees. Many propagation phenomena in real life adhere to the three-degree influence principle, such as behaviors related to obesity, smoking, alcohol consumption, and the spread of emotions on the internet [142]. Furthermore, the three-degree influence principle can be used to determine the range of measurement for key node mining algorithms based on local topology [143,144]; hence, the step length for the three-degree decay Grover walk algorithm discussed in this section is set to 3.

According to the three-degree influence principle, as the levels of propagation deepen, the effectiveness of influence propagation will diminish.

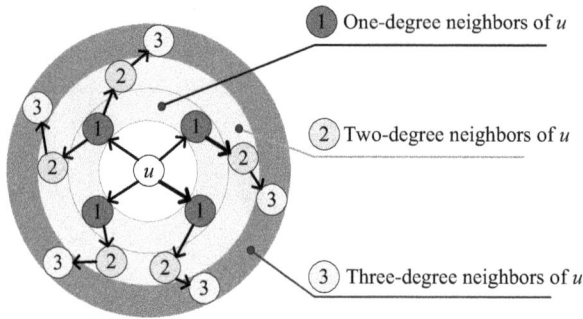

FIGURE 3.3   Illustration of the three-degree of influence principle.

Let the attenuation coefficient be β, and the measurement probability of node j at the tth step of the walk be $P(j;t)$. Then the importance degree $Sg(j)$ of node j in the network is defined as follows:

$$Sg(j) = \beta \times P(j;1) + (1-\beta) \times \sum_{t=2}^{3} P(j;t). \tag{3.25}$$

Based on the three-degree influence principle, the influence propagation in the second and third degrees also experiences attenuation, so it is viewed as a whole multiplied by the coefficient $(1-\beta)$. Obviously Eq. (3.25) can effectively express the characteristics of the three-degree influence principle only when $\beta > 1/2$; otherwise, there is no attenuation effect for second and third-degree propagation. Moreover, we can identify the range of parameter β, where $\beta \in (0.5, 0.9)$. In this section, we let $\beta = (0.5 + 0.9)/2 = 0.7$. In Eq. (3.25), the measurement probability of node j at step t is computed using the method referenced in Eq. (2.6), defined as

$$P(j;t) = \sum_{k=1}^{N(j)} \left| \psi_{j,k}(t) \right|^2, \tag{3.26}$$

where the measurement result of node j is the sum of the amplitudes of node j and its neighboring nodes. Equation (3.26) pertains to the measurement results at the node level, while the measurement results for all nodes in the network can be expressed as a probability distribution. Let $P(t)$ represent the probability distribution obtained from measuring all nodes after t

steps of the walk. For the entire network system, the importance scores of all nodes can be conveyed as $P = \beta \times P(t=1) + (1-\beta) \times \sum_{\varepsilon=2}^{3} P(t=\varepsilon)$, this expression suggests the measurement method at the network level.

3.2.3.2 Key Node Mining Using the Three-Degree Decay Grover Walk Algorithm

To validate the performance of the proposed three-degree decay Grover walk algorithm in key node mining tasks, the Chesapeake, Adjnoun, Enron-only, and Jazz networks were selected as experimental datasets, with introductions to the four network datasets provided in the appendix of this book. This section analyzes the effectiveness of the three-degree decay Grover walk algorithm in key node mining tasks using the SIR model, Kendall coefficient, and robustness index as evaluation criteria. The principles and interpretations of these three evaluation metrics are sourced from Section 3.1. The experiments chose degree centrality (DC) [137], betweenness centrality (BC) [137], closeness centrality (CC) [137], PageRank [145], local triangle centrality (LTC) [146], k-shell [147], and QPageRank [119] algorithms for comparison. Figure 3.4 depicts scatter

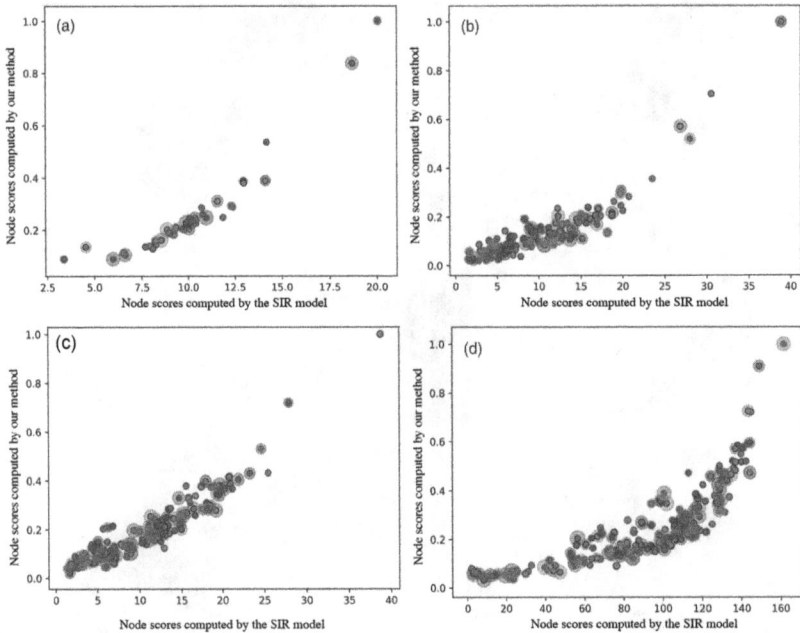

FIGURE 3.4 Experiment on influence consistency under the SIR model. (a) Chesapeake. (b) Adjnoun. (c) Enron-only. (d) Jazz.

plots of the impact measurement values of all nodes in the network using the three-degree decay Grover walk algorithm and the SIR model, representing the experimental results of impact consistency between the two methods. Each scatter point is assigned a random size (radius) and a random color. The measurement results of the SIR model are often used as a standard to assess the performance of algorithms in ranking key nodes. It is evident that the SIR model and the three-degree decay Grover walk algorithm exhibit extremely strong consistency in measuring nodes across the four network datasets, directly reflecting the accuracy of the three-degree decay Grover walk algorithm in ranking key nodes in the network.

The measurement results of the SIR model serve only as a reference. To demonstrate the performance of the introduced three-degree decay Grover walk algorithm from multiple perspectives as comprehensively as possible, the correlation of measurement results of various algorithms for network nodes is quantified based on the Kendall coefficient in Eq. (3.2). The experimental results are presented in Figure 3.5, where each small

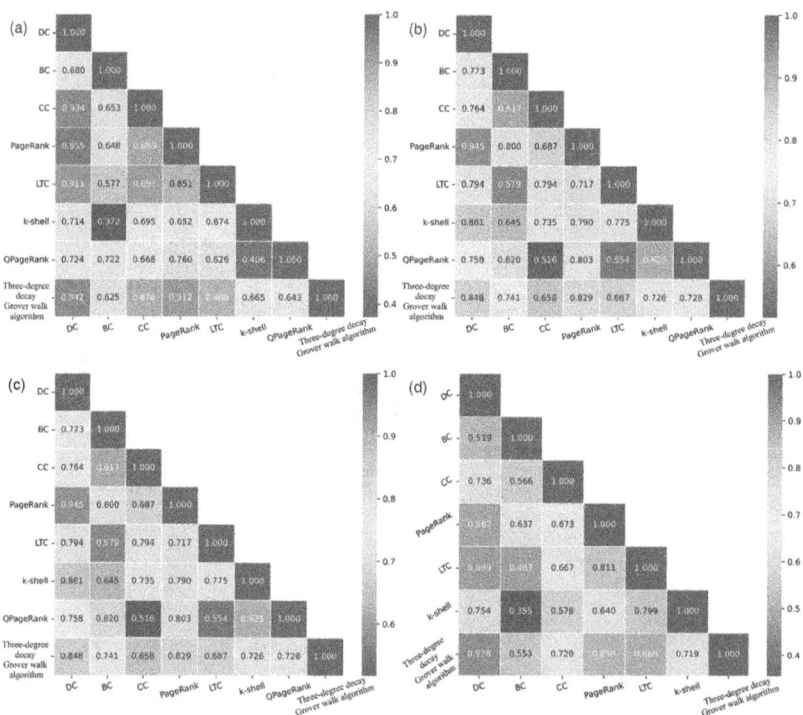

FIGURE 3.5 Key node mining based on the Kendall coefficient. (a) Chesapeake. (b) Adjnoun. (c) Enron-only. (d) Jazz.

box represents the Kendall coefficient of two algorithms. In Figure 3.5, the average Kendall coefficient of the DC method and the three-degree decay Grover walk algorithm is as high as 0.906, and the average Kendall coefficient of the LTC algorithm and the three-degree decay Grover walk algorithm is 0.798. These results indicate that the three-step walk accumulation results used by the three-degree decay Grover walk algorithm effectively contain both degree values and structural information of triadic closures, which are beneficial for accurately measuring key nodes in the network. Compared with the global random walk-based PageRank algorithm, the average Kendall coefficient with the three-degree decay Grover walk algorithm is 0.844, but the walk steps of the three-degree decay Grover walk algorithm amounts to only 3. Similarly, as a DTQW algorithm with high complexity in simulation on existing computers, the QPageRank algorithm cannot obtain measurement results for key nodes in the Enron-only and Jazz networks within a reasonable time. In contrast, the three-degree decay Grover walk algorithm exhibits better scalability in practical applications than the well-known QPageRank algorithm.

Finally, based on the robustness index in Eq. (3.1), Figure 3.6 shows the destruction ability of key nodes ranked by various algorithms on

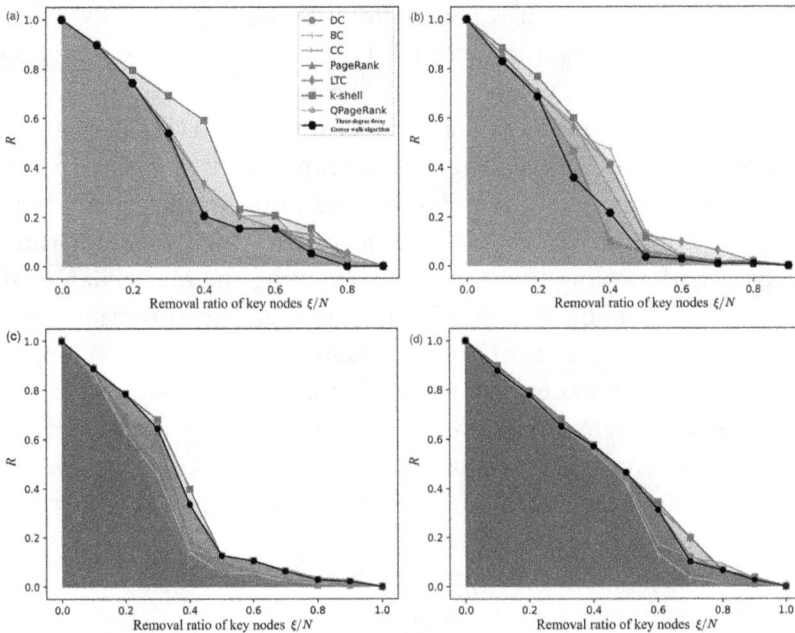

FIGURE 3.6   Key node mining based on the robustness index. (a) Chesapeake. (b) Adjnoun. (c) Enron-only. (d) Jazz.

network stability. When removing nodes in the sequence by proportion, the more apparent the network disintegration caused by the removal (the smaller the R value), the more important the removed nodes are. According to Figure 3.6, the three-degree decay Grover walk algorithm demonstrates good ranking performance in the networks corresponding to (a), (b), and (d). Notably, in the Adjnoun network, the key nodes selected by the three-degree decay Grover walk algorithm can quickly dismantle the original network into several independent subgraphs.

3.3 CONTINUOUS-TIME QUANTUM WALKS IN NODE DISCOVERY

This section introduces two continuous-time quantum walk (CTQW) algorithms and an information propagation model based on CTQW, all of which are utilized to mine the key nodes of complex networks.

3.3.1 Google PageRank Algorithm for Open Quantum System

The open quantum system PageRank algorithm, discussed in Section 3.2.1, is based on CTQWs. This algorithm was published in the journal *Scientific Reports* in 2012 [131], which represents a pioneering work in using CTQWs for mining key nodes in complex networks. Due to its focus on open systems, it is hereinafter referred to as the OPR (open-system PageRank) algorithm. In the OPR algorithm, an open system specifically refers to a scenario where the adjacency matrix relevant to a directed complex network is a non-Hermitian matrix. The motion of particles on the network nodes cannot be expressed by a positive matrix; the system is considered a non-isolated quantum system (open system).

The scoring process of the OPR algorithm for node importance involves the following three steps: (1) Defining the Markovian quantum master equation in the form of quantum measurement, which replaces the traditional Schrödinger equation in CTQWs; (2) Using the Lindblad equation to describe the motion of particles within the complex network system; (3) Converting the nonlinear Lindblad differential equation into a linear equation through a linearization method, and using the measurement results of each node as the importance score of that node.

The first step of the OPR algorithm is the definition of the Markovian quantum master equation. To begin with, the classical Markov random process is defined as

$$\frac{d}{dt}p_i = \sum_j M_{ij}p_j, \tag{3.27}$$

where M is the difference between the probability transition matrix and the identity matrix, that is, $M = P^G - \hat{I}$. For reference on the probability transition matrix, see Eq. (3.6) in Section 3.2.1. Furthermore, using $\langle i|\psi \rangle$ to denote probability p_i and $\langle i|H|j \rangle$ to represent matrix M, and introducing a Schrödinger equation in the form of Eq. (2.15) to rewrite Eq. (3.27), the following equation can be obtained:

$$\frac{d}{dt}\langle i|\psi \rangle = -\frac{i}{\hbar}\sum_j \langle i|H|j \rangle \langle j|\psi \rangle. \tag{3.28}$$

In quantum mechanics, an arbitrary wave function can be substituted by a density operator. Let the density operator be ϱ and define it as

$$\varrho = \sum_i p_i |\psi_i \rangle \langle \psi_i|. \tag{3.29}$$

In CTQWs, the evolution of the subsystem is provided by the Schrödinger equation. For open quantum systems, consider using the Lindblad equation to describe the evolution of particles at network nodes. By applying density operators to process Eq. (3.28), the following can be acquired:

$$\frac{d\varrho}{dt} = \mathcal{L}\varrho, \tag{3.30}$$

where $\mathcal{L}$ is the differential operator. The differential operator in any Markov master equation can be expressed as

$$\mathcal{L}_\varrho = -i(1-\alpha)[H,\varrho] + \alpha \sum_{(i,i)} \gamma_{(i,j)}\left(L_{(i,j)}\varrho L^\dagger_{(i,j)} - \frac{1}{2}\{L^\dagger_{(i,j)}L_{(i,j)},\varrho\} \right), \tag{3.31}$$

where i and i are used to distinguish between subscripts and imaginary numbers; H is the symmetrization of the network adjacency matrix, $H = H^\dagger$, and γ is essentially the adjacency matrix of the network, which is typically explained as a special case of the Google matrix [148]. Equations (3.30) and (3.31) provide an irreversible dynamical method that can be used to achieve particle transitions at directed network nodes.

Let $L_{(i,j)} = |i\rangle\langle j|$ and express the density operator ϱ in diagonal matrix form. When α takes different values in Eq. (3.31), the following equations can be inferred:

$$\begin{cases} \dfrac{d}{dt}\varrho_{ii} = \displaystyle\sum_j \left(\gamma_{ij} - \delta_{ij}\right)\rho_{jj}, & \alpha = 1 \\[2em] \dfrac{d}{dt}\varrho_{ii} = -i\displaystyle\sum_j 2H_{ij}\left(\varrho_{ji} - \varrho_{ij}\right), & \alpha = 0 \end{cases} \tag{3.32}$$

Finally, the probability of the particle remaining at node j is specified as the projection of the standard basis of node j on the density operator:

$$P(j) = \langle j|\varrho|j \rangle = \varrho_{jj}. \tag{3.33}$$

where $P(j)$ is the score assigned by the OPR algorithm to the importance of node j in network G.

To demonstrate the performance of the OPR algorithm in the task of ranking key nodes, we take the network shown in Figure 3.7a as an example and choose the classic PageRank and random walk algorithms for comparison. The ranking results for the nodes in the network of Figure 3.7a are illustrated in Figure 3.7b. In Figure 3.7b, the horizontal coordinates represent the ranking sequence of the nodes according to the random walk algorithm. It can be observed that in this numbering order, the PageRank algorithm yields results completely consistent with those of the random walk algorithm, while the PageRank algorithm only differs from the OPR algorithm in the ranking positions of nodes 5 and 7. The experiments displayed in Figure 3.7 prove that the OPR algorithm not

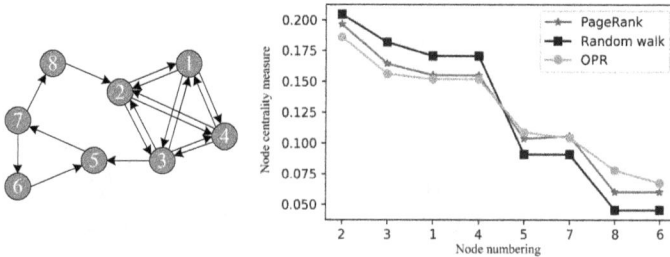

FIGURE 3.7 Network data and experimental outcomes for the OPR algorithm. (a) OPR algorithm network data. (b) Node centrality measurement.

only realizes quantum walks in open systems but also accurately ranks the importance of nodes in complex networks.

It is worth noting that due to the presence of numerous Kronecker operations, the simulation of the OPR algorithm on existing computers incurs significant costs [121]. In 2021, Tang et al. addressed this issue by parallelizing the computation process of the OPR algorithm based on TensorFlow [149], effectively reducing computational costs and making the OPR algorithm scalable to larger complex networks.

3.3.2 Quantum Jensen-Shannon Divergence Centrality Algorithm

The quantum Jensen-Shannon divergence (QJSD) algorithm was defined by Rossi et al. [150], and the main idea of this algorithm for mining central nodes in a network is as follows: First, a CTQW is given based on the normalized Laplacian matrix of the graph; then, the QJSD is designed using the density operator formula of the von Neumann entropy and the classical Jensen-Shannon divergence. Finally, the initial probability amplitudes of the same node are represented with opposite phase signs, resulting in two density operators for a single node. These two distributions are used as inputs for QJSD, and the QJSD metric value serves as the centrality score of the node.

For a complex network G, the QJSD algorithm first replaces the Hamiltonian H in the Schrödinger equation with the normalized Laplacian matrix L of the graph. In accordance with Eqs. (2.15) and (2.17), the evolution equation of the QJSD algorithm can be described as

$$|\psi(t)\rangle = e^{-iLt}|\psi(0)\rangle, \tag{3.34}$$

where $|\psi(0)\rangle = \sum_{j=1}^{N} \alpha_j(0)$ indicates the quantum state at the initial time, where $\alpha_j(0)$ denotes the probability amplitude of node j at the initial moment. Since the Laplacian matrix L can be decomposed into a product of diagonal matrices, let the eigenvector matrix of L be Φ, where $\Phi = (\phi_1, \phi_2, \cdots, \phi_N)$, and the diagonal matrix of L be Λ, where $\Lambda = \text{diag}(\lambda_1, \lambda_2, \cdots, \lambda_N)$ and $0 = \lambda_1 \leq \lambda_2 \leq \cdots \leq \lambda_N$. Therefore, Eq. (3.34) can be expressed as

$$|\psi(t)\rangle = \Phi e^{-i\Lambda t}|\psi(0)\rangle. \tag{3.35}$$

In the next step, based on the von Neumann entropy and classical distribution similarity calculation methods, the scoring formula for QJSD on key nodes is determined. In quantum mechanics, the density operator ρ can describe a pure state $|\psi_i\rangle$ of a quantum system [19], where each state corresponds to a probability p_i, and ρ is defined as

$$\rho = \sum_i p_i |\psi_i\rangle\langle\psi_i|. \tag{3.36}$$

Furthermore, let ξ_i represent the eigenvalues of the density operator ρ, where $\xi_i = \{\xi_1, \xi_2, \cdots, \xi_N\}$. The von Neumann entropy of the density operator ρ is defined as

$$H_N = -\mathrm{tr}(\rho \log \rho) = -\sum_i \xi_i \ln \xi_i. \tag{3.37}$$

In the calculation result of Eq. (3.37), if the system represented by the density operator is a mixed state (mixed state quantum system), the result of H_N is not equal to 0; conversely, if the system is a pure state (pure state quantum system), the result of the von Neumann entropy H_N is equal to 0 [19].

Established on the above definitions, and combining the JS divergence, the QJSD can be designed. The classical JS divergence is used to measure the similarity between two probability distributions and can be considered a variant of the KL divergence (Kullback–Leibler divergence), improving the nonsymmetric metric issue of the KL divergence. For any two probability distributions ρ and σ, the classical JS divergence is formulated as

$$\mathrm{JS}(\rho\|\sigma) = \frac{1}{2}\mathrm{KL}\left(\rho\left\|\frac{\rho+\sigma}{2}\right.\right) + \frac{1}{2}\mathrm{KL}\left(\sigma\left\|\frac{\rho+\sigma}{2}\right.\right). \tag{3.38}$$

If the calculation method of KL( ) in Eq. (3.38) is substituted with the von Neumann approach from Eq. (3.37), a new metric formula for JS divergence is generated:

$$D_{\mathrm{JS}}(\rho,\sigma) = H_N\left(\frac{\rho+\sigma}{2}\right) - \frac{1}{2}H_N(\rho) - \frac{1}{2}H_N(\sigma), \tag{3.39}$$

where $D_{\mathrm{JS}}(\rho,\sigma) \in [0,1]$.

Finally, by taking the opposite sign of the initial probability amplitude of any node j in the network, two density operators ρ_{j-} and ρ_{j+} associated with node j are constructed. Equation (3.39) is then employed to score the centrality of node j. The specific calculation process is as follows. Assign $\left|\psi(0)^{-}\right\rangle = \sum_{j \in V} \alpha_j^{-}(0)|j\rangle$ and $\left|\psi(0)^{+}\right\rangle = \sum_{j \in V} \alpha_j^{+}(0)|j\rangle$, and let ρ_{j-} and ρ_{j+} represent the density operators of quantum states $\left|\psi(t)^{j-}\right\rangle$ and $\left|\psi(t)^{j+}\right\rangle$, respectively, calculated as

$$
\begin{cases}
\rho_{j-} = \lim_{T \to \infty} \frac{1}{T} \int_0^T \left|\psi(t)^{j-}\right\rangle\left\langle\psi(t)^{j-}\right| dt \\
\rho_{j+} = \lim_{T \to \infty} \frac{1}{T} \int_0^T \left|\psi(t)^{j+}\right\rangle\left\langle\psi(t)^{j+}\right| dt
\end{cases}
\tag{3.40}
$$

At this point, by substituting the two density operators corresponding to node j from Eq. (3.40) into Eq. (3.39), we obtain the centrality score of node j:

$$
C_{QJSD}(j) = D_{JS}\left(\rho_{j-}, \rho_{j+}\right).
\tag{3.41}
$$

Figure 3.8 shows the test network for the QJSD algorithm, which describes the marital relationships between 16 socially prominent families [151]. Each box relates to a family surname. The QJSD algorithm's scoring results for the nodes in this network are presented in Table 3.2. Combining Figure 3.8 and Table 3.2, the QJSD algorithm's ranking results for this

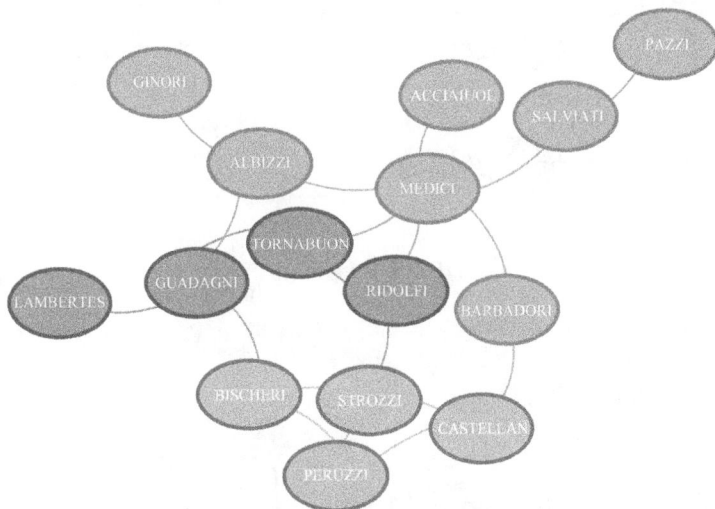

FIGURE 3.8   Test network for the QJSD algorithm.

TABLE 3.2   QJSD Sorting Results of Nodes in the Family Network [150]

Families	Scores	Families	Scores	Families	Scores
MEDICI	0.4867	CASTELLAN	0.3245	SALVIATI	0.2248
RIDOLFI	0.4619	BARBADORI	0.3205	GINORI	0.1993
STROZZI	0.4192	ALBIZZI	0.3172	ACCIAIUOL	0.1534
TORNABUON	0.4041	GUADAGNI	0.3091	LAMBERTES	0.1267
BISCHERI	0.3586	PERUZZI	0.2990	PAZZI	0.1126

network align with the meaning of the network data. For instance, MEDICI, RIDOLFI, and STROZZI, which are in the core positions of the network topology, rank in the top three, while GINORI, ACCIAIUOL, LAMBERTES, and PAZZI, which are in the peripheral positions, have the lowest centrality scores and rank at the bottom.

The above experiment intuitively demonstrates the effectiveness of the QJSD algorithm in identifying key nodes in the network. This effectiveness stems from the significant correlation between the QJSD algorithm's calculation results and degree centrality. Taking two density operators ρ_{j-} and ρ_{j+} corresponding to node j as an example, the proof process is as follows:

$$D_{JS}\left(\rho_{j-},\rho_{j+}\right)=H_N\left(\frac{\rho_{j-}+\rho_{j+}}{2}\right)-\frac{1}{2}H_N\left(\rho_{j-}\right)$$

$$=-\frac{\mu_0+1}{2}\log_2\frac{\mu_0+1}{2}-\sum_{i\neq 0}\frac{\mu_i}{2}\log_2\frac{\mu_i}{2}+\frac{1}{2}\sum_i\mu_i\log_2\mu_i$$

$$=\frac{\mu_0+1}{2}-\frac{\mu_0+1}{2}\log_2\left(\mu_0+1\right)+\sum_{i\neq 0}\frac{\mu_i}{2}-\frac{1}{2}\sum_{i\neq 0}\mu_i\log_2\mu_i$$

$$+\frac{1}{2}\sum_i\mu_i\log_2\mu_i$$

$$=1-\frac{1}{2}\log_2\left(\mu_0+1\right)+\frac{\mu_0}{2}\log_2\frac{\mu_0}{\mu_0+1}. \tag{3.42}$$

The calculation method for μ_0 is given by

$$\mu_0=\left\langle\phi_0|\rho_0|\phi_0\right\rangle=\left\langle\phi_0|\psi_0^{j-}\right\rangle^2=\left(1-\frac{|N(j)|}{|E|}\right)^2. \tag{3.43}$$

In Eq. (3.43), $N(j)$ signifies the set of the nearest neighbor nodes of the node, and $|E|$ denotes the total number of network links. According to Eqs. (3.42) and (3.43), it can be observed that the QJSD algorithm's measurement results are closely related to the degree centrality of the nodes. The QJSD algorithm has become a crucial step in solving graph network similarity problems [152–154].

3.3.3 Information Propagation Model Based on Quantum Walks

This section introduces a continuous-time quantum walk-based information propagation (CTQW-IP) model [155]. The model evaluates the influence of a node in a network by simulating the influence propagation range of a specified node. Experiments reveal that the model can effectively rank high-influence nodes in a social network.

3.3.3.1 Definition of CTQW-IP Model

In the CTQW-IP model, the temporal evolution of the entire network system matches with the process of information dissemination. The setting of the initial probability amplitude for an individual node designates that node as the seed (initial propagation node), while the quantum measurement phase serves as a statistical method for the results of information propagation. This is the core idea of using CTQWs to develop information propagation models.

Unlike existing information propagation models, the CTQW-IP model maps all nodes in the social network one-to-one with elements in the quantum state, denoted as $|\psi\rangle$. After specifying the seed node, the evolution operator provided by the Schrödinger equation is used to simulate the process of information transfer in the social network through the transfer of probability amplitudes between nodes. Finally, by setting a predefined threshold and utilizing the quantum measurement process, the propagation results of the initial seed nodes are statistically analyzed, and the importance score of nodes in the network is evaluated. Based on the above process, the description of the CTQW-IP model in this section is categorized as: specifying the seed node through preset probability amplitudes, simulating information propagation using quantum walk evolution, and calculating the influence of seeds through quantum measurement, as shown in Figure 3.9.

First, the seed node is specified by presetting probability amplitudes. Figure 3.9a provides an example, where the red ball is the initial seed for information propagation. In the CTQW-IP model, any node can be designated as a seed by presetting the initial probability amplitude, which is the

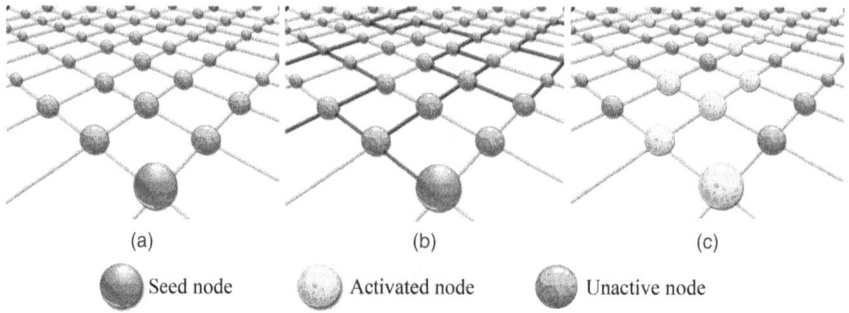

(a)   (b)   (c)

Seed node   Activated node   Unactive node

FIGURE 3.9 Three core steps of the information dissemination model [155]. (a) Assigned seed node. (b) Simulated information dissemination. (c) Statistical propagation results.

initial source of information propagation in the network. The task of the CTQW-IP model is to evaluate its influence in the network by simulating information propagation. Referring to the definition form of CTQWs in Section 2.1.2, the quantum state on the network G at the initial moment is expressed as

$$|\psi(0)\rangle = \sum_{j \in V} \alpha_j(0)|j\rangle, \tag{3.44}$$

where $\alpha_j(0)$ is the probability amplitude of node j at the initial moment: $|j\rangle$ is the standard basis pertaining to node j. Let $N(j)$ denote the set of nearest neighbors of node j, then the probability amplitude presetting method for specifying the seed node is given by

$$\begin{cases} \alpha_k(0) = \dfrac{1}{\sqrt{|N(j)|}}, & k \in N(j) \\ \alpha_v(0) = 0, & v \in V \setminus N(j) \end{cases} \tag{3.45}$$

The meaning of Eq. (3.45) is that at the initial moment, the probability amplitudes of the neighboring nodes of the seed j are evenly distributed, while the probability amplitudes of other nodes are set to 0, thereby designating the seed and specifying the propagation path of the seed at the initial moment.

After that, simulate information propagation is simulated using quantum walk evolution. The evolution process of the CTQW-IP model relates to the information transmission process in the classical cascading information propagation model. In this model, the evolution dynamics are provided by the Schrödinger equation:

$$i\frac{d}{dt}|\psi(t)\rangle = H|\psi(t)\rangle, \tag{3.46}$$

where H represents the Hamiltonian of the system, which can be replaced by the adjacency matrix of the social network graph. For any undirected social network G, its adjacency matrix can be defined as

$$A_{i,j} = \begin{cases} 1, & (i,j) \in E \\ 0, & (i,j) \notin E \end{cases}. \tag{3.47}$$

Substituting Eq. (3.47) into the Hamiltonian H in Eq. (3.46), the solution of Eq. (3.47) can be obtained:

$$|\psi(t)\rangle = \exp(-iAt)|\psi(0)\rangle \tag{3.48}$$

As shown in Figure 3.9b, information starts from the seed node and can only propagate in a cascading manner (marked in red) depending on the connectivity of the network [139], and cannot be transmitted without the link relationship.

Quantum measurement is the final step of the CTQW-IP model, used to calculate the influence of the seed node after evolution simulation on the social network. As a result, the influence of seed j will be calculated using the quantum walk measurement formula, and its calculation method is

$$\begin{aligned} P(t,j) &= \left|\langle j|\psi(t)\rangle\right|^2 \\ &= \left|\langle j|\exp(-iAt)|\psi(0)\rangle\right|^2 . \end{aligned} \tag{3.49}$$

Directly using Eq. (3.49) to evaluate the influence of the seed node will cause the following problems: (1) The results obtained by this measurement method are only related to the network topology and evolution characteristics, which have no relation to whether the nodes in the propagation results are in an activated state. (2) After calculating Eq. (3.49) for each node in the network, a probability distribution will be obtained. Reasonably utilizing this probability distribution to evaluate the influence of the seed node will become an essential part of the CTQW-IP model design.

Through the above analysis, this section outlines a threshold method, considering nodes with a probability distribution greater than the threshold Θ as activated nodes. The more activated nodes, the greater the influence of the seed. This threshold method is described as

$$\text{Inf}(u) = \sum_{i=1}^{N} Q(i), \tag{3.50}$$

where the function $Q(\bullet)$ is characterized as

$$Q(i) = \begin{cases} 1, & P(t,i) > \Theta \\ 0, & \text{otherwise} \end{cases}. \tag{3.51}$$

The threshold Θ of the CTQW-IP model is set as the reciprocal of the number of network nodes, $\Theta = N^{-1}$, where N^{-1} represents the average probability of the particle staying at each node. Through quantum measurement, when the probability of the particle staying at a certain node is greater than the average probability N^{-1}, it can be considered that the node is an activated node of the seed. For example, in the propagation results shown in Figure 3.9c, the yellow individuals indicate nodes that are in an activated state after the propagation ends. It is notable that the threshold parameter in the CTQW-IP model is different from the threshold parameter in the linear threshold (LT) model [139]. In the LT model, the threshold determines whether the current node is activated and whether information can cascade transmit, while the threshold Θ in the CTQW-IP model does not participate in the cascading transmission process of information, but is only used to statistically analyze the propagation results of seed nodes.

3.3.3.2 Parameter Determination and Experimental Verification of CTQW-IP Model

Due to the unknown time parameter t in the CTQW-IP model, the measurement results of Eq. (3.49), when transformed using Euler's formula, yield a complex trigonometric function, leading to significant iterative computational costs and the issue of non-convergence in propagation results [132]. To address this, predetermining the parameter t in the CTQW-IP model can simplify the calculations and prevent the propagation results from being time-variant.

This section determines an appropriate value for the parameter t in advance by observing how the probability distribution results change

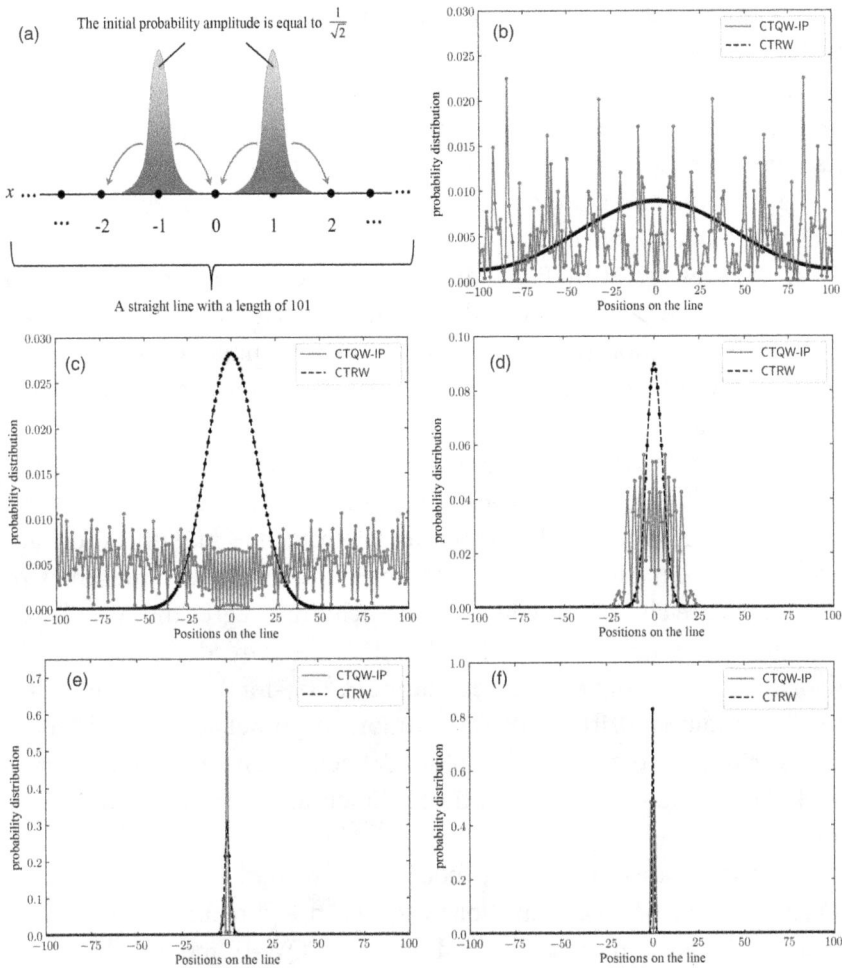

FIGURE 3.10   Probability distribution of the CTQW-IP model for various values of t [155]. (a) The initial probability amplitude. (b) $t=10^3$. (c) $t=10^2$. (d) $t=10^1$. (e) $t=10^0$. (f) $t=10-1$.

with variations in t. Considering a one-dimensional line of length 101 as shown in Figure 3.10a, for example, the central node at $x=0$ is set as the initial position for particle walk, and the initial probability amplitudes are set according to Eq. (3.45). When the parameter t takes values of 10^3, 10^2, 10^1, 10^0, and 10^{-1}, the probability distribution results are presented in Figure 3.10b–f. In Figure 3.10, for ease of observing the performance of the CTQW-IP model on a line, a classical continuous-time random walk (CTRW) is included for comparison.

In Figure 3.10b and c, the measurement probabilities at the two ends of the line are significantly higher than the probability at the central node $x = 0$. Clearly, the values of t taken as 10^3 and 10^2 are not suitable for identifying highly influential nodes in the network. Similarly, when t is 10, it also does not meet the measurement requirements for the central node. In Figure 3.10e and f, although both results largely align with the probability distribution of classical random walks, when $t = 10^{-1}$, the probability amplitudes preset by Eq. (3.45) play a more pronounced role, which helps ensure the accuracy of seed influence calculations by specifying the path for information transmission from the seed node to its neighbors. Therefore, in this section's experiments, the parameter t for the CTQW-IP model is set to 10^{-1}.

This section further validates the influence of the seed nodes selected by the CTQW-IP model through the influence maximization experiment. The definition and description of influence maximization are referenced from Section 3.1. In this experiment, five representative algorithms are chosen for comparison, including degree discount [134], strongly connected component (SCC) [156], newGreedy [157], a hybrid algorithm combining degree descending & lazy-forward (DDLF) [158], and a random algorithm. Three common social networks are selected as experimental data: Infect-dublin, Caltech36, and Hamsterster networks, with their statistical indicators mentioned in the appendix of this book.

The performance of the seed nodes chosen by the CTQW-IP model under the influence maximization is shown in Figure 3.11. According to Figure 3.11, the seed nodes selected by the CTQW-IP model achieve the largest propagation scale in the network, reflecting the accuracy of the CTQW-IP model in selecting seed nodes. Additionally, taking the DDLF algorithm as an example, in the influence maximization experiment results of Figure 3.11c, the DDLF algorithm only outperforms the random algorithm; while in the experiment results of the network in Figure 3.11d, DDLF ranks just below the CTQW-IP model suggested in this section, indicating that the DDLF algorithm is sensitive to propagation parameters. In contrast, the CTQW-IP model not only identifies highly influential nodes in social networks but also maintains stable performance in influence maximization when the propagation probability parameters take different values.

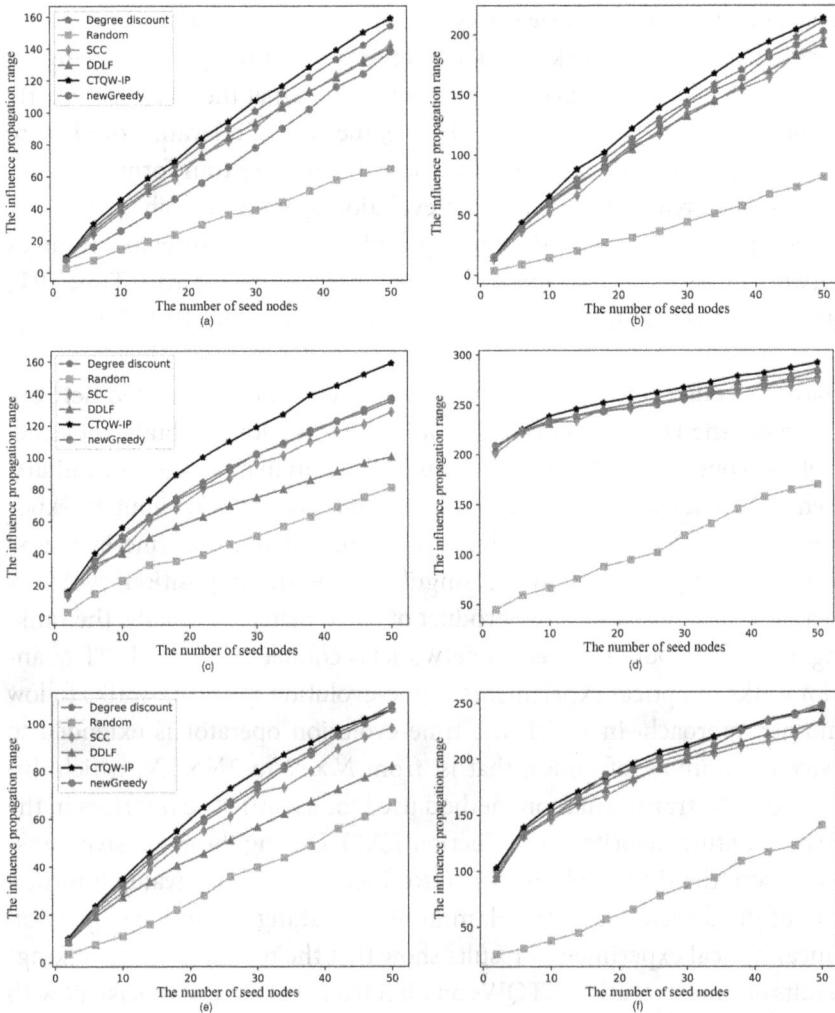

FIGURE 3.11 Propagation outcomes of the influence maximization [155]. (a) $p=0.01$, Infect-dublin. (b) $p=0.05$, Infect-dublin. (c) $p=0.01$, Caltech36. (d) $p=0.05$, Caltech36. (e) $p=0.01$, Hamsterster. (f) $p=0.05$, Hamsterster.

3.4 SUMMARY AND EXTENSIONS

This chapter introduces six ranking algorithms for key nodes in complex networks from both DTQWs and CTQWs, primarily focusing on undirected complex networks and their simulation effects on existing computers. Other research on key node mining in complex networks based on quantum walks contrasts with the above approach, as researchers focus on directed networks under the condition of non-Hermitian matrices, exploring how to achieve reversible particle walks on directed graphs, that is, simulating a directed complex network through quantum walks [159]. The challenge of quantum walks on directed complex networks is that the adjacency matrix of the network is non-Hermitian. Hence, the core of relevant work is constructing an evolution operator that meets the unitary transformation conditions, often referred to as the time-evolution operator in such studies.

Two representative works on key node mining in directed complex networks based on quantum walks are presented here: Parity-Time (PT) quantum walks [160] and time-evolution quantum walks [161]. The design concept of PT quantum walks is as follows: First, a pseudo-Hermitian matrix is constructed from the adjacency matrix of the directed network to replace the Hamiltonian in Schrödinger's equation to obtain the time-evolution operator. Although the pseudo-Hermitian matrix is non-unitary, it ensures that the total probability amplitude oscillates without an exponential surge or decrease during evolution. Second, the time-evolution operator is diagonalized using singular value decomposition, which is further decomposed into the product of three matrices. Finally, the ranking of central nodes in directed networks is completed through PT quantum walks in optical experiments. Time-evolution quantum walks follow another approach, in which the time-evolution operator is extended to twice the number of nodes, that is, from $N \times N$ to $2N \times 2N$, which differs from the transformation method used for asymmetric matrices in the HHL quantum algorithm (see Section 1.2.3). During the expansive transformation, the time-evolution operator incorporates eigenvalue information of the directed network Hamiltonian, making it a unitary operator. Linear optical experimental results show that the measurement (ranking) results of time-evolution CTQWs on directed networks are consistent with the node rankings obtained by the PageRank algorithm. Since the relevant results focus on the physical implementation of optical experiments, which goes beyond the scope of algorithm design and implementation, this chapter does not provide detailed introduction.

The focus of this chapter is on node discovery. While numerous research findings on network representation learning indicate that the complexity of complex networks (irregular characteristics) primarily depends on links [162]. Hypothetically, in an extreme scenario where a complex network only contains nodes without links, the visualization results would be a set of scattered points, lacking complexity. Therefore, the objects of network structure mining are not limited to nodes but can also include links bridging two nodes. Chapter 4 will introduce quantum walk algorithms for link mining in complex networks.

Applications of Quantum Walks in Link Mining

4.1 LINK MINING: DEFINITION AND EVALUATION METHODS

Based on diverse research objectives, link mining tasks in complex networks can be divided into two categories: mining critical links in complex networks and mining potential or missing links, the latter often referred to as link prediction. Below, two brief examples are provided to illustrate the practical value of critical link mining and predicting missing links in complex networks in real-life scenarios. Assume a network is under attack from an unknown third party. To prevent a cascading failure across the entire network, certain central nodes within the network can be removed. However, if this network represents a transactional network constructed between banks and merchants, the removal of the key node (i.e., the bank) would disrupt the entire network's operation. In such a situation, severing some critical links to prevent a network-wide cascade might allow the network to continue functioning normally. A similar example can be seen in complex networks composed of servers and clients, where the server acts as a central node and cannot be easily removed from the network. In such cases, cutting critical links connected to the server node, when necessary, may be a viable option. Clearly, mining critical links in complex networks is of significant research importance for ensuring network reliability.

The task of link prediction involves inferring the missing or potentially future links within a network, which are not equivalent to the critical links

DOI: 10.1201/9781003683902-6

of the network. The essence of link prediction is to forecast the evolutionary trends of the network based on the similarity of the topological structures of nodes. This approach has substantial benefits, particularly in the field of biology. For instance, in the protein metabolism network, the interactions between proteins are surmised based on numerous experimental reactions rather than directly detected through an electron microscope. Observing the interactions between proteins in the microscopic world requires significant investments in reagents, equipment, time, and human resources. Even so, our understanding of biological macromolecules remains superficial. For example, as of 2008, only 20% of all interactions between yeast proteins were known [163]. Link prediction can infer relationships between macromolecules based on known topological structures, serving as theoretical guidance for biological experiments. This method reduces the economic costs and time required for deducing protein topologies, providing tangible benefits for human science and health.

This chapter will focus on the two tasks of critical link mining and link prediction, introducing the application of quantum walks in link mining within complex networks.

4.1.1 Critical Link Mining and Its Evaluation Metrics

Critical links play vital roles across various fields. To illustrate, adding critical links to a network can promote information dissemination in social networks [164]; critical links can measure the quality of network connectivity [165]; and critical links can also be used to explain the role of phase transitions in percolation networks [166]. The difficulty of mining critical links varies among different types of networks. In Figure 4.1a, for instance, the network exhibits a clear community structure, where the link marked with a triangle is evidently a critical link for the entire network. If this link is severed, the entire network will no longer be connected. Another type of network is shown in Figure 4.1b, which lacks obvious topological features, rendering it difficult to identify critical links. In practical scenarios, most complex networks follow self-organizing or power-law distribution properties, imposing complications to clearly define communities and hierarchies. Therefore, the task of mining critical links remains a significant challenge.

Whether a link is considered critical is generally determined using network robustness metrics. In simple terms, robustness is assessed by removing a specific link from the network and calculating the extent to which the network disintegrates as a result. The degree of disintegration

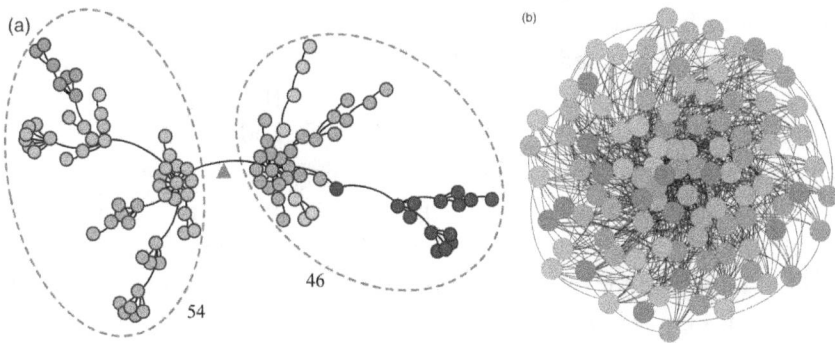

FIGURE 4.1 Critical link mining in different complex networks [169]. (a) Easy to identify the critical link. (b) Difficult to identify the critical link.

is expressed as the ratio of the number of nodes in the largest connected component of the network after the link is removed to the total number of nodes in the network. Let σ_ζ represent the ratio of the number of nodes in the largest connected component to the total number of nodes after the removal of the ζth link, where the ζth link is selected based on its score assigned by the algorithm, with higher-scoring links being removed first. Let the complex network be denoted as $G = (V, E)$, where V and E signify the sets of nodes and links, respectively, with $|V| = N$ and $|E| = M$. The robustness metric for the ζth link is then defined as follows:

$$\eta = \frac{1}{M} \sum_{\zeta=1}^{M} \sigma_\zeta, \tag{4.1}$$

where $1/M$ normalizes the measurement result, so $\eta \in [0,1]$. Illustrating with the network shown in Figure 4.1a, which consists of 100 nodes and 99 links, when the critical link marked with a triangle is severed, the network is divided into two separate connected components with 54 and 46 nodes, respectively. Therefore, according to Eq. (4.1), the importance of this link in the network can be quantified as $(1/99) \times (54/100) = 0.00545$. This chapter uses this metric to evaluate the performance of algorithms in mining critical links. A lower η value indicates a higher degree of network damage after removing the link, suggesting that the link is more critical.

4.1.2 Link Prediction and Its Evaluation Metrics

Assume that with a probability $\mu \in (0,1)$, the set of links E in network G is randomly classified into a training link set E^T and a prediction link set E^P.

The training link set E^{T} is utilized to store the training data, while the prediction link set E^{P} is used to record the missing links in the network, which are the links that the algorithm needs to predict. The relationship between these two sets satisfies $|E^{\mathrm{T}}| = \mu |E|$, $E^{\mathrm{T}} \cap E^{\mathrm{P}} = \varnothing$, and $E^{\mathrm{T}} \cup E^{\mathrm{P}} = E$. For the complex network $\tilde{G}$ composed of the node set V and the training link set E^{T}, $\tilde{G} = (V, E^{\mathrm{T}})$, the objective of the link prediction task is to score all the links in the network using an algorithm to predict the links in set E^{P}. Here, all the links stand for the set of links in the complete graph formed by all nodes in V, denoted as E^{U}, where $E \subseteq E^{\mathrm{U}}$. In other words, when performing link prediction with an algorithm, it is essentially scoring the links of the complete graph corresponding to network $\tilde{G}$, rather than only the links existing in network $\tilde{G}$ [167]. Figure 4.2 illustrates an example of the application of link prediction in social networks. Suppose two users both have accounts on Weibo and LinkedIn. These two users follow each other on Weibo but have not established a connection on LinkedIn. Based on the similarity of their social relationship on Weibo, it is possible to predict that they may form a friendship on LinkedIn in the future, adding a link between them to make them mutually connected friends. Thus, recommendation systems perform as a highly intuitive application scenario for link prediction.

The main basis for link prediction is the similarity information between nodes [168], and the accuracy of the algorithm's prediction results also depends on the scoring of node similarities. This chapter uses two metrics to validate the advantages of the suggested simplified

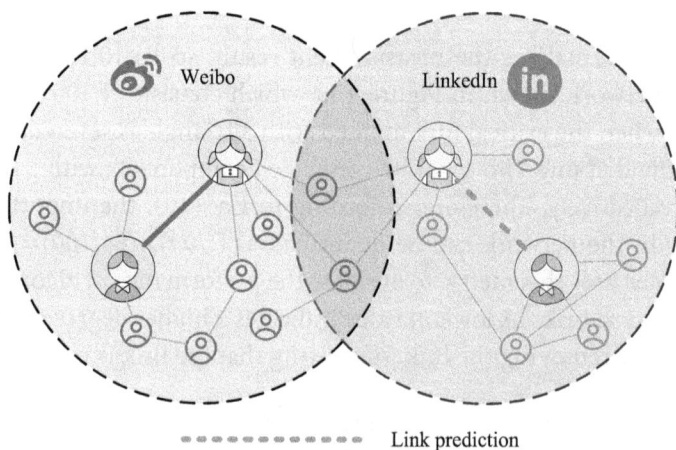

FIGURE 4.2   Link prediction with application to social networks [169].

quantum walk algorithm in link prediction, including the area under the receiver operating characteristic curve (AUC) and the precision metric. In particular, when different algorithms have similar values under the AUC metric, the precision metric can further determine which algorithm performs better.

For network $\tilde{G}$, the AUC metric is calculated as follows: A sample is taken from both the training link set E^{T} and the non-existent link set $(E^{\mathrm{U}} - E)$. If the score of the link from the training link set E^{T} is greater than the score of the link from the non-existent link set $(E^{\mathrm{U}} - E)$, 1 point is recorded; if the two scores are equal, 0.5 points are recorded; otherwise, 0 points are recorded. Let the number of samples be n. The formula for calculating the AUC metric is as follows:

$$\mathrm{AUC} = \frac{n_1 + 0.5 n_2}{n}, \tag{4.2}$$

where $n = n_1 + n_2$. An algorithm with a higher AUC value indicates more accurate predictions of missing links in the network. It should be emphasized that when the number of samples n is set to 672,400, the absolute error of the AUC calculation can be guaranteed not to exceed 0.001 with 90% confidence, regardless of the network's size and characteristics. The detailed proof process can be found in Appendix A of *Link Prediction* [168]. The precision metric records the proportion of correctly predicted links among the top L links. Let l represent the number of correctly predicted links; the precision metric is then given by

$$\mathrm{Pre} = \frac{l}{L}. \tag{4.3}$$

When the AUC values of various algorithms are similar, the algorithm with a higher precision metric demonstrates a clearer advantage in link prediction.

4.2 QUANTUM WALKS IN CRITICAL LINK IDENTIFICATION

This section presents a quantum walk algorithm driven by the Hadamard coin operator [169], hereinafter referred to as the Hadamard walk algorithm, which can accurately mine critical links in complex networks. Generally, the Hadamard operator, Grover operator, Fourier operator, and SU(2) operator [17] can all function as coin operators for discrete-time quantum walks. However, both the Grover and Fourier operators are biased operators. Biased operators are not suitable for mining critical links; particularly, the relationship between the measurement results of the Fourier operator and the network's topological characteristics is quite ambiguous and requires further exploration. When the two phase parameters in the SU(2) operator are set to 0 and the angle parameter is set to $\pi/4$, the SU(2) operator becomes equivalent to the Hadamard operator [36]. On the other hand, as an unbiased coin, the Hadamard operator drives particles to walk on the network solely based on the existing link relationships, allowing it to accurately reflect the network's topology and further facilitate the mining of critical links in complex networks. Accordingly, this section considers using the Hadamard operator as the coin to drive the quantum walk on complex networks.

4.2.1 Hadamard Walk Algorithm on a Static Complex Network

The Hadamard walk algorithm discussed in this section is described based on the general framework outlined in Section 2.3. Assume that the Hadamard walk occurs on an undirected complex network G. Since the dimensions of the total Hilbert space and the matrix size of the evolution operator must be equal, this section considers using the direct sum operation to accumulate the space in relation to each node. The corresponding spatial dimension is defined as follows:

$$\mathcal{H} = \mathcal{H}_1 \oplus \mathcal{H}_2 \oplus \cdots \oplus \mathcal{H}_N, \tag{4.4}$$

where $\mathcal{H}$ is composed of the Hilbert spaces $\mathcal{H}_j$ associated with the node, where $\forall j \in V$. The dimension of the Hilbert space $\mathcal{H}_j$ is equal to the number of nearest neighbors of node j, denoted as $|N(j)|$. Hence, the dimension of the total Hilbert space $\mathcal{H}$, denoted as $D_{\mathcal{H}}$, is defined as follows:

$$D_{\mathcal{H}} = \sum_{j=1}^{N} |N(j)|. \tag{4.5}$$

Apparently, $D_{\mathcal{H}}$ is the sum of the degrees of all nodes in the network, which is twice the number of links in the network, that is, $D_{\mathcal{H}} = 2M$. The second part of the Hadamard walk algorithm is the state vector. At the initial moment, the state vector is formulated as:

$$|\psi(0)\rangle = \sum_{j=1}^{N} \sum_{k=1}^{N(j)} \alpha_{j,k}(0)|j,k\rangle$$

(4.6)

$$= \frac{1}{\sqrt{N}} \sum_{j=1}^{N} \sum_{k=1}^{N(j)} \frac{1}{\sqrt{|N(j)|}} |j,k\rangle,$$

where $\alpha_{j,k}(0)$ denotes the probability amplitude of the particle transitioning from node k to node j at the initial moment, and $\alpha_{j,k}(0) \in [0, 1]$; the transition of the particle from node j to node k is expressed as a standard basis $|j,k\rangle$. The sum of the squared moduli of the probability amplitudes on the complex network remains equal to 1, and its formal expression can be found in Eq. (3.21) of Section 3.2.3.

The coin operator of the Hadamard walk algorithm is the Hadamard operator, and its multiplication with the shift operator constitutes the evolution operator of the Hadamard walk algorithm. Let S denote the shift operator and H symbolize the Hadamard operator. The evolution operator U is defined as follows:

$$U = S(H \otimes \hat{i}),$$

(4.7)

where $\hat{i}$ is the identity matrix; the expression for matrix H can be found in Eq. (1.11). The shift operator S functions similarly to a flip-flop between a pair of nodes. Thus, the shift operator S is described as

$$S|j,k\rangle = |k,j\rangle.$$

(4.8)

Each step of the Hadamard walk algorithm aligns to one application of the evolution operator U. Therefore, after t steps, the evolution of the state vector is illustrated as

$$|\psi(t)\rangle = U^t |\psi(0)\rangle$$

$$= U|\psi(t-1)\rangle.$$

(4.9)

The purpose of the measurement phase is to score the nodes at both ends of a link, which is the final step of the Hadamard walk algorithm. Since a link connects two nodes, it is naturally intuitive to use a two-particle Hadamard walk algorithm to score each link in the network. This is because if the two particles co-occur at the two ends of a link with a high probability during the measurement, it indicates that the link is more important. However, if the two-particle Hadamard walk algorithm is used to mine the critical links in the network, the evolution operator in the Hadamard walk algorithm needs to be expanded to accommodate the movement of two particles between network nodes. For example, the shift operator $\hat{S}$ in the two-particle Hadamard walk algorithm would be identified as

$$\hat{S}|j,k;p,q\rangle=|k,j\rangle\otimes|q,p\rangle, \tag{4.10}$$

where $|k,j\rangle$ indicates that one particle transitions from node j to node k; $|q,p\rangle$ shows that another particle transitions from node p to node q. According to the properties of the tensor product, the matrix size of the shift operator $\hat{S}$ increases from $2M$ to $4M^2$. Consequently, the dimensions of the Hilbert space and the state vector also increase. With clarity, the two-particle Hadamard walk algorithm requires significant computational resources for mining critical links in the network, especially in large-scale networks, compromising its feasibility.

In fact, if the two particles do not interact, the two-particle Hadamard walk algorithm can be replaced by the single-particle Hadamard walk algorithm. In other words, for any link $E(j,k)\in E$, the joint measurement result of nodes j and k using the two-particle Hadamard walk is equivalent to the composite result of the independent measurements of nodes j and k using the single-particle Hadamard walk algorithm. When using the single-particle Hadamard walk algorithm, the dimension of the shift operator remains $2M$, without consuming additional computational resources. Hence, the scoring method for a link $E(j,k)$ using the Hadamard walk algorithm is specified as follows:

$$P_{E(j,k)}^{(t)} = e^{P(j;t)}e^{P(k;t)}, \tag{4.11}$$

$$P(j;t)= \sum_{k=1}^{N(j)}\left|\alpha_{j,k}(t)\right|^2, \tag{4.12}$$

where $P_{E(j,k)}^{(t)}$ denotes the scoring result of the link $E(j,k)$ after t steps; $P(j;t)$ and $P(k;t)$ are the measurement results for nodes j and k, respectively, after t steps using the single-particle Hadamard walk algorithm. In Eq. (4.11), since the measurement results are decimal values between 0 and 1, $P(j;t) \in [0, 1]$, an exponential function e^x is introduced to convert all decimal values for the nodes into values greater than 1. For this reason, the larger the value of $P_{E(j,k)}^{(t)}$, the more important the link $E(j,k)$ is in the network. The calculation methods for $P(j;t)$ and $P(k;t)$ can be found in Eq. (4.12).

4.2.2 Critical Link Identification Using the Hadamard Walk Algorithm

To validate the effectiveness of the Hadamard walk algorithm highlighted in this section for critical link mining, two experiments are designed. These experiments demonstrate the algorithm's average performance and detailed performance in critical link mining, with both experiments using Eq. (4.1) as the evaluation metric. In these experiments, the Dolphins, Polbooks, Adinoun, Jazz, Metabolic, and Email networks are selected as the datasets. The Fourier walk algorithm [111], Grover walk algorithm [111], QPageRank algorithm [119], Degree Product (DP) algorithm [170], and Diffusion Intensity (DI) algorithm [171] are employed for comparison. The statistical characteristics and descriptions of the experimental network data are available in the appendix of this book.

4.2.2.1 Average Performance of Critical Link Mining

The average performance of the algorithm in mining critical links refers to the mean value of each link calculated using Eq. (4.1). It reflects the overall effectiveness of the algorithm in solving the critical link mining problem. Since the performance of the algorithm is evaluated based on the extent of network disintegration after removing critical links, a lower metric value in the average performance experiment indicates better algorithm performance. The experimental results of different algorithms in terms of average performance in critical link mining are shown in Figure 4.3. Each square represents the overall performance of an algorithm in mining critical links within a specific network. The lighter the color of the square, the better the critical link mining performance of the corresponding algorithm.

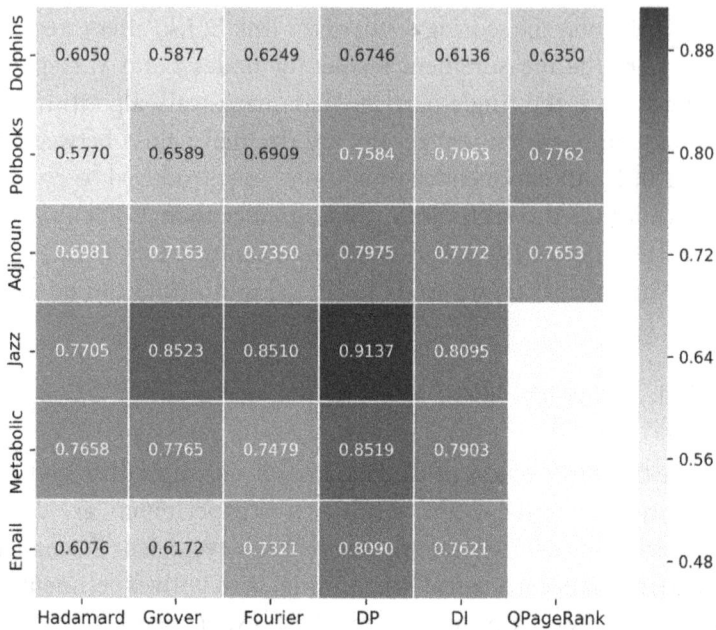

FIGURE 4.3 Average performance of different algorithms in critical link mining [169].

Based on the experimental results in Figure 4.3, the following analysis is provided: (1) For any network dataset, quantum walk-based algorithms (Hadamard walk algorithm, Grover walk algorithm, Fourier walk algorithm, and QPageRank algorithm) demonstrate superior critical link mining capabilities compared to the DI and DP algorithms. Except for the Dolphins network, where the average performance of the Grover walk algorithm surpassed the Hadamard walk algorithm suggested in this chapter, the Hadamard walk algorithm achieved the best average performance among the compared algorithms in the other five networks. (2) Due to spatial complexity constraints, the average performance of the QPageRank algorithm could not be calculated within an effective time for the Jazz, Metabolic, and Email networks. Although quantum walk-based algorithms exhibit substantial advantages in critical link mining, the QPageRank algorithm is less applicable across networks of different scales in comparison with the other three coin-based quantum walk algorithms

(Hadamard walk algorithm, Grover walk algorithm, and Fourier walk algorithm).

Furthermore, the related improvement percentage (RIP) is introduced to quantify the accuracy differences between the Hadamard Walk algorithm and other algorithms in critical link mining within complex networks. The RIP is defined as $\overline{RIP} = \sum_{i=1}^{\kappa} \eta_{others}^i / \eta_H^i - 1$, where η_{others}^i signifies the robustness metric value of a comparison algorithm in the ith network dataset, η_H^i denotes the robustness metric result of the Hadamard walk algorithm in the same network, and κ indicates the number of networks used by the current algorithm. According to the $\overline{RIP}$, when comparing the Grover walk algorithm, Fourier walk algorithm, and QPageRank algorithm for identifying critical links in networks, the accuracy of the Hadamard walk algorithm discussed in this chapter is relatively improved by 4.59%, 9.49%, and 15.55%, respectively. In contrast to the non-quantum walk algorithms, DP and DI, the accuracy of the Hadamard walk algorithm for identifying critical links is comparatively improved by 20.03% and 11.48%, respectively. In summary, the accuracy of the Hadamard walk algorithm in critical link mining is improved by 4.59%–20.03% relative to other algorithms, indicating that the Hadamard walk algorithm is capable of accurately identifying critical links in complex networks.

4.2.2.2 Detailed Performance of Critical Link Mining

The overall performance of various algorithms in critical link mining is still evaluated using Eq. (4.1). This experiment refines the importance of each link ranked by the algorithms. In addition, the detailed performance of different algorithms in critical link mining is presented in Figure 4.4, where the black solid line represents the Hadamard walk algorithm, and the comparison algorithms are depicted as colored dashed lines. According to the robustness metric from Eq. (4.1), when all links in the network are removed, the robustness measure will inevitably be 0, meaning that each algorithm's curve in Figure 4.4 must be a declining curve from 1 to 0. A smaller robustness metric indicates that the link is more critical. As such, the rate of decline in the curves is a crucial piece of information in the experimental results shown in Figure 4.4. Moreover, the criticality of links pertains only to the small number of top-ranked links, not all links. This concept is aligned with studies on maximizing influence in social networks [134] and identifying key nodes in complex networks [172]. Therefore, in the analysis of experimental results in Figure 4.4, the

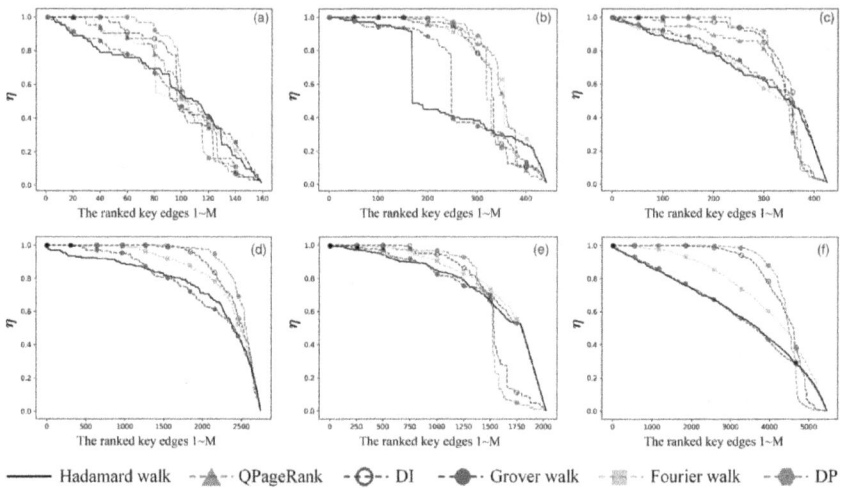

FIGURE 4.4 Detailed performance of different algorithms in critical link mining [169]. (a) Dolphins. (b) Polbooks. (c) Adjnoum. (d) Jazz. (e) Metabolic. (f) Email.

focus is on examining the rate of decline in the algorithm's curve after deleting the top-half (top-50%) of the links.

On the basis of the experimental results in Figure 4.4, the following analysis is provided: (1) In the critical link mining results for the first half of the links, the Hadamard walk algorithm reveals the fastest rate of decline in its corresponding curve, which is particularly evident in Figure 4.4b–e. (2) The ranking of the critical link mining capabilities among the three coin-based quantum walk algorithms, from strongest to weakest, is as follows: Hadamard walk algorithm > Grover walk algorithm > Fourier walk algorithm. (3) In Figure 4.4e and f, although the DI and DP algorithms perform more accurately in ranking the links in the latter half, these links are non-critical. The critical link mining problem concentrates only on the critical minority within the network. Therefore, it can be concluded that the Hadamard walk algorithm demonstrates high accuracy in identifying critical links in complex networks.

4.2.3 Hadamard Walk in a Dynamic UAV Network

In recent years, UAVs (unmanned aerial vehicles) have played a significant role in wildlife conservation, environmental monitoring, aerial photography, and forest fire prevention [173]. Assuming UAVs are represented as nodes and the communication relationships between them as links, the communication status of a UAV swarm can be modeled as a complex network within a unit plane. Particularly, due to the limitation of transmission power, UAVs can only communicate within a certain range, making the UAV communication network a typical temporal complex network. This section considers a dynamic network using the UAV communication network as an example and applies the Hadamard walk algorithm to identify critical UAV nodes within the dynamic network.

This section includes: defining the UAV model used to simulate the dynamic UAV communication network; analyzing the principles of using the Hadamard walk to mine critical UAV nodes; and comparing with other classical algorithms to evaluate the effectiveness of the Hadamard walk algorithm in identifying key nodes in dynamic networks.

4.2.3.1 Simulation Model of UAV Communication Network

Building on complex network theory, this section describes a simulation model for the UAV communication network. As shown in Figure 4.5a, UAVs are represented as nodes, and effective communication relationships between UAVs are abstracted as links. The communication status of UAVs at any given time forms a complex network composed of nodes and links. It is assumed that the UAVs' flight status is displayed within a unit plane, and the flight state of any UAV is determined by three controllable parameters: flight speed v_i, variable flight angle ϕ_i, and relative flight position Pos_i. At time τ, the relative flight position Pos_i of UAV i in the unit plane is identified by its flight angle ϕ_i. This process is illustrated in Figure 4.5b, and its formal expression is given by

$$Pos_i^{(\tau)} = \left(\cos\phi_i^{(\tau)}, \sin\phi_i^{(\tau)} \right). \tag{4.13}$$

Based on the definition in Eq. (4.13), when the relative positions of all UAVs are known, the Euclidean distance formula can be used to determine

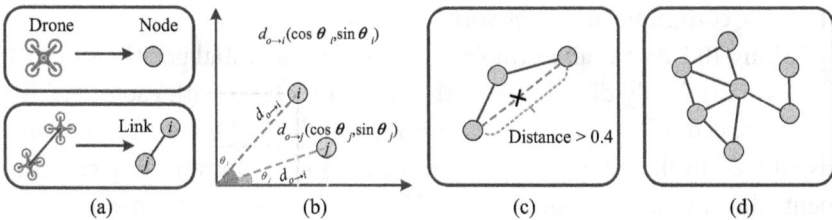

FIGURE 4.5 Simulation process of the UAV communication network [169].

the communication distance between any pair of UAVs. In the unit plane, let $d_{i,j}^{(\tau)}$ represent the straight-line distance between UAV i and UAV j. The calculation formula is specified as follows:

$$d_{i,j}^{(\tau)} = \sqrt{\left(\cos\phi_i^{(\tau)} - \cos\phi_j^{(\tau)}\right)^2 + \left(\sin\phi_i^{(\tau)} - \sin\phi_j^{(\tau)}\right)^2}. \qquad (4.14)$$

If a valid communication radius γ is given, then a link can form between UAV i and UAV j if and only if the straight-line distance between them is less than the radius γ; otherwise, no link exists between them.

Using the effective communication radius and the straight-line distance between UAVs to determine whether a link exists between nodes is analogous to the concept of an adjacency matrix composed of 0 and 1. Let $E^{(\tau)}$ represent the set of links in the UAV communication network at time τ. The process of constructing the communication relationships between UAVs at time τ can then be described as follows:

$$\begin{cases} E(i,j) \in E^{(\tau)}, & d_{i,j}^{(\tau)} \leq \gamma \\ E(i,j) \notin E^{(\tau)}, & d_{i,j}^{(\tau)} > \gamma \end{cases}. \qquad (4.15)$$

In general, the transmission power of UAVs within a swarm is equal, so there is only one effective communication radius globally. In this experiment, γ is fixed at 0.4. Figure 4.5c provides a simple example of the calculation process described above. In summary, the model used to simulate the UAV communication network can be characterized as

$$\Im = \{\tau, \gamma, N, E^{(\tau)}, v, \phi, \text{Pos}\}. \qquad (4.16)$$

At any given time, a network snapshot can be obtained, consisting of a set of UAV nodes N and a set of links $E^{(\tau)}$. Figure 4.5d showcases an example of a UAV communication network snapshot.

Taking 10 UAVs as an example, assume that the initial positions of each UAV are randomly distributed within the unit plane, with each UAV flying at a speed of 0.02 units per time unit, and the flight angle ϕ uniformly distributed in the interval $(0, 2\pi)$. In the LabView programming environment, the simulation results of the UAV communication network snapshots at equal time intervals are depicted in Figure 4.6.

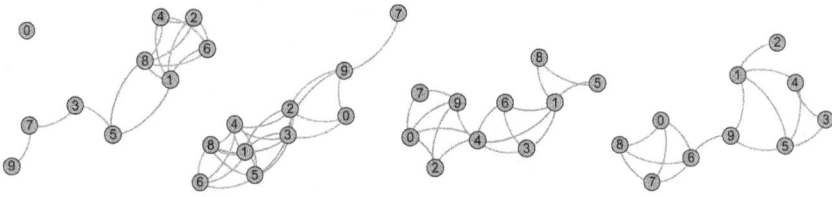

FIGURE 4.6 Snapshots of UAV communication networks at the same time interval [169].

4.2.3.2 The Principle of Hadamard Walk in Mining Key Nodes of Dynamic Networks

In dynamic complex networks, a link may exist in the network snapshot at the current moment but may disappear in the next moment. As part of the critical links, UAV nodes in dynamic UAV networks do not vanish; only the communication relationships, that is, the links between the UAV s, may change. Critical links function as the primary basis for mining key nodes in dynamic networks. Moreover, since the measurement results of the Hadamard walk algorithm are not directly related to key nodes, this section utilizes the frequency of node occurrences within critical links to determine the key UAV nodes in the dynamic UAV communication network. UAV nodes with higher occurrence frequencies are more important within the entire communication network.

The process of using the Hadamard walk algorithm for identifying key nodes in dynamic networks can be referenced in Figure 4.7. First, the dynamic UAV communication network in flight is expressed as network snapshots at different time points, with time as the sequence. The dynamic network is presented in Figure 4.7a. Taking the network snapshot at a specific time point, as shown in Figure 4.7b, the Hadamard walk algorithm is applied to rank the criticality of the 23 links within the network snapshot from high to low, with the ranking results displayed in Figure 4.7c. According to the experiments in Section 4.2.2, under robustness metrics, the Hadamard walk algorithm exhibits extremely high accuracy in ranking the top-50% of critical links. Therefore, the top-50% of the ranked results in Figure 4.7c are selected as the candidate set for critical UAV nodes. This result is rounded down, leading to the outcome in Figure 4.7d. Each link contains two nodes; hence, the occurrence frequency of all nodes within the top-50% of critical links is recorded, and this frequency is used as the

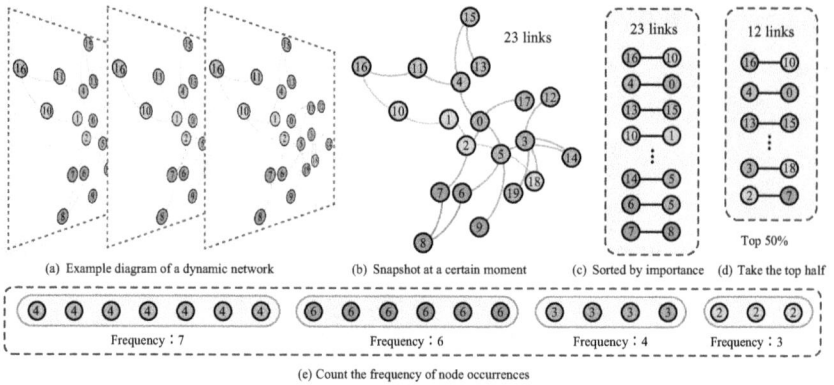

(a) Example diagram of a dynamic network   (b) Snapshot at a certain moment   (c) Sorted by importance   (d) Take the top half

(e) Count the frequency of node occurrences

FIGURE 4.7   Flowchart of the Hadamard walk algorithm for identifying key nodes in dynamic networks [169].

score for assessing the importance of the UAV nodes. Figure 4.7e illustrates a possible ranking result.

The Hadamard walk algorithm is utilized to identify key UAV nodes in dynamic UAV communication networks based on the occurrence frequency of nodes within the top-50% of critical links. Unlike traditional algorithms for identifying central nodes in dynamic networks, this approach of determining key nodes based on critical links takes into account the connectivity, reachable paths, and node frequency information of dynamic complex networks. This represents an extended application of the Hadamard walk algorithm in dynamic networks.

4.2.3.3 Experiment on Mining Key Nodes in Dynamic Complex Networks

This section selects the temporal degree centrality (TDC) algorithm [174], temporal PageRank (TPR) algorithm [145], and temporal betweenness centrality (TBC) algorithm [175] as comparison methods. The snapshots of the UAV communication network shown in Figure 4.6 are used as experimental data for the dynamic network. The number of key nodes is set to 5, and the experimental results of key node mining in the dynamic network are presented in Table 4.1. In the table, the metric values of the key nodes selected by each algorithm across diverse snapshots are calculated based on the SIR model, with each metric value being the average of 10^4 repeated calculations. The method of evaluating key nodes using the SIR model can be found in Section 3.1.

TABLE 4.1   Results of Key Node Mining in Dynamic Networks

Algorithms	Snapshot 1	Snapshot 2	Snapshot 3	Snapshot 4	Results	Node Number
Hadamard	**6.233**	6.530	5.654	**7.688**	**26.105**	0,4,6,7,8
TDC	5.552	6.478	5.830	6.096	23.956	1,4,5,6,8
TPR	5.573	6.506	5.840	6.123	24.042	1,4,5,6,8
TBC	6.084	7.301	6.269	5.759	25.413	1,4,5,6,9

Note: Bolded data indicates the maximum value of the simulation results across different snapshots.

Built upon the experimental results presented in Table 4.1, it is evident that the Hadamard walk algorithm discussed in this chapter outperforms the other comparison algorithms in the task of mining key nodes in dynamic networks, achieving the highest cumulative value in the final simulation results based on the SIR model. The TDC algorithm is primarily influenced by two factors when selecting key nodes: (1) The degree values of nodes may vary significantly across different time points; (2) In dense network snapshots, the degree values of most nodes are equal. Therefore, it is challenging for the TDC algorithm to effectively identify key nodes in dynamic complex networks. As for the TPR algorithm, it is mainly designed for directed complex networks, which limits its effectiveness in undirected dynamic UAV communication networks. Additionally, although both the TBC and Hadamard walk algorithms are sensitive to the time complexity of dense networks, the Hadamard walk algorithm demonstrates superior performance in identifying key nodes in dynamic complex networks compared to the TBC algorithm.

Accurately identifying key UAV nodes in dynamic UAV communication networks enables the implementation of special protection measures for these critical UAVs, thereby enhancing the robustness of the UAV communication network. Thus, the Hadamard walk algorithm suggested in this chapter demonstrates practical value in the structural analysis of complex networks.

4.3 QUANTUM WALKS IN LINK PREDICTION

This section introduces a link prediction algorithm based on continuous-time quantum walks and a discrete-time quantum walk algorithm incorporating a simplification approach. Both algorithms can accurately predict missing links in complex networks.

4.3.1 Quantum Link Prediction Algorithm

The quantum link prediction (QLP) algorithm is currently the only known link prediction algorithm that utilizes continuous-time quantum walks, which was defined by Moutinho et al. [176]. Based on the definition of the continuous-time quantum walk evolution equation in Eq. (2.18) of Section 2.1.2, it is evident that when the Hamiltonian in the Schrödinger equation is substituted by the adjacency matrix of the network, the evolution operator constructed from e^{-iAt} evolves based on network connectivity. This computation process can be expanded using the Taylor equation:

$$e^{-iAt} = \sum_{n=0}^{k} \frac{1}{n!}(-iAt)^n.$$ (4.17)

where A^n represents the matrix where the path length between any two nodes is equal to n. When n is an odd number $(n = 2k+1)$, the result of Eq. (4.17) contains an imaginary component; when n is even $(n = 2k)$, the result contains no imaginary component. In light of this, the QLP algorithm utilizes the odd and even path lengths of continuous-time quantum walks to predict possible or missing links within the network. Furthermore, the QLP algorithm contains the preparation of quantum states, the construction of evolution operators, and the definition of the quantum measurement process based on the odd and even powers. Figure 4.8 displays the quantum circuit diagram of this algorithm.

According to Figure 4.8, the qubits in the QLP algorithm are divided into two parts: the first part represents N nodes using n qubits, where $n = \log N$; the second part consists of auxiliary qubits $|c\rangle_a$. Therefore, the quantum state of node j at time t is captured as

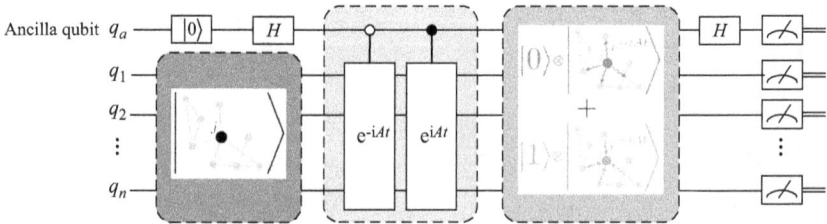

FIGURE 4.8 Quantum circuit diagram of the quantum link prediction algorithm [176].

$$\left|\psi_j(t)\right\rangle = \sum_{c=0}^{1}\left|c\right\rangle_a\left(\frac{e^{-iAt}+(-1)^c\,e^{iAt}}{2}\right)\left|j\right\rangle_n. \tag{4.18}$$

To make the relationship between the odd-even concept of the QLP algorithm and link prediction more explicit, Eq. (4.18) is rewritten as follows:

$$\left|\psi_j(t)\right\rangle = \left|0\right\rangle_a\left(\sum_{k=0}^{+\infty}c_{\text{even}}(k,t)A^{2k}\right)\left|j\right\rangle_n + i\left|1\right\rangle_a\left(\sum_{k=0}^{+\infty}c_{\text{odd}}(k,t)A^{2k+1}\right)\left|j\right\rangle_n, \tag{4.19}$$

where $c_{\text{odd}}(k,t)$ and $c_{\text{even}}(k,t)$ are time-independent coefficients defined for odd and even cases, respectively. Their calculation methods are given by

$$c_{\text{odd}}(k,t) = \frac{(-1)^{k+1}\,t^{2k+1}}{(2k+1)!},\quad c_{\text{even}}(k,t) = \frac{(-1)^k\,t^{2k}}{(2k)!}. \tag{4.20}$$

For the initial node j, the probability of a particle residing on node i is distinguished and measured separately based on odd and even cases. Let $p_{i,j}$ denote the prediction score for the link $e(i,j) \in E$. The calculation method for $p_{i,j}$ is presented as

$$p_{i,j}^{\text{odd}} \propto \left|\left\langle i\right|\left(\sum_{k=0}^{+\infty}c_{\text{odd}}(k,t)A^{2k+1}\right)\left|j\right\rangle\right|^2. \tag{4.21}$$

$$p_{i,j}^{\text{even}} \propto \left|\left\langle i\right|\left(\sum_{k=0}^{+\infty}c_{\text{even}}(k,t)A^{2k}\right)\left|j\right\rangle\right|^2. \tag{4.22}$$

In the quantum circuit, the prediction score $p_{i,j}$ for the link $e(i,j)$ cannot be directly read out. However, the QLP algorithm can repeatedly sample the measured distribution results to identify the link $e(i,j)$ whose measurement probability is proportional to $p_{i,j}^{\text{even}}$ or $p_{i,j}^{\text{odd}}$.

Built on the QLP algorithm, the prediction performance is generally and significantly better when the path length is odd compared to when it is even. In 2021, Zhou et al. explored the impact of 2-hop and 3-hop path lengths on link prediction accuracy across 137 complex networks [177].

A large number of experiments, based on the AUC and precision metrics, demonstrated that both path lengths can accomplish the link prediction task. However, algorithms based on 3-hop paths exhibited more advantages in low-density networks with low average clustering coefficients. When the path lengths are odd and even, the features they contain correspond to all nodes' information from the 3-hop and 2-hop paths, respectively. Since the QLP algorithm indicates that link prediction performs better with odd path lengths, this section focuses on analyzing the similarities and differences between the QLP algorithm and the well-known 3-hop link prediction algorithms. The experiments used Yeast, Facebook, and Wiki-vote networks as the experimental data; thorough descriptions of these networks are provided in the appendix of this book. For comparison, the linear optimization (LO) algorithm [178], the third-order adjacency matrix of the network (A^3), and the degree-normalized paths of length three (L3) algorithm [179] were selected. The predicted scores for all links in the network from different algorithms were sorted in descending order, and the top 10^4 links were taken as samples. The overlap ratio of the link prediction results between the QLP algorithm and LO, A3, and L3 algorithms is presented in Table 4.2. For instance, in the Yeast network, when the parameter t of the QLP algorithm is set to 0.1, the QLP algorithm and the LO algorithm share 8,911 identical links among the top 10^4, which is equivalent to 89.11%. In Table 4.2, the parameter t in the second column specifically refers to the time parameter in the QLP algorithm, and the information in parentheses indicates the value of parameter α in the LO algorithm.

Drawing from the experimental results in Table 4.2, the following conclusions can be reached: (1) As the parameter t in the QLP algorithm increases, the link prediction accuracy of the QLP algorithm decreases compared to other algorithms based on 3-hop neighborhood information. (2) When the t value of the QLP algorithm is relatively small, the prediction results of the QLP algorithm are highly consistent with those of other 3-hop link prediction algorithms. The experiments demonstrate that when the path length is odd and the parameter t is small, the QLP algorithm, founded on continuous-time quantum walks, can accurately predict the potential links in complex networks with high precision.

4.3.2 Simplified Quantum Walk Algorithm

The simplified quantum walk algorithm is an achievement by the author of this book, published in the journal *Entropy* [180]. The simplification

TABLE 4.2   Overlap Ratios of Link Prediction Results between QLP Algorithm and LO, A^3, and L3 Algorithms

Networks	t value	QLP with LO	QLP with A^3 (%)	QLP with L3 (%)
Yeast	0.001	100.00% ($\alpha = 1.67 \times 10^{-10}$)	99.72	89.42
	0.010	100.00% ($\alpha = 1.67 \times 10^{-7}$)	99.72	89.41
	0.050	97.17% ($\alpha = 2.08 \times 10^{-5}$)	96.64	89.35
	0.100	89.11% ($\alpha = 1.67 \times 10^{-4}$)	85.18	89.11
	0.500	40.98% ($\alpha = 2.08 \times 10^{-2}$)	10.95	81.55
	0.800	44.60% ($\alpha = 8.53 \times 10^{-2}$)	5.67	67.15
	1.000	47.53% ($\alpha = 1.67 \times 10^{-1}$)	3.36	51.13
Facebook	0.001	99.98% ($\alpha = 1.67 \times 10^{-10}$)	99.98	97.58
	0.010	98.77% ($\alpha = 1.67 \times 10^{-7}$)	98.74	97.58
	0.050	59.75% ($\alpha = 2.08 \times 10^{-5}$)	55.79	97.64
	0.100	50.40% ($\alpha = 1.67 \times 10^{-4}$)	41.87	97.55
	0.500	28.27% ($\alpha = 2.08 \times 10^{-2}$)	50.30	54.47
	0.800	24.10% ($\alpha = 8.53 \times 10^{-2}$)	31.70	30.66
	1.000	23.11% ($\alpha = 1.67 \times 10^{-1}$)	27.20	36.92
Wiki-vote	0.001	100.00% ($\alpha = 1.67 \times 10^{-10}$)	99.99	94.35
	0.010	98.46% ($\alpha = 1.67 \times 10^{-7}$)	98.41	94.34
	0.050	53.63% ($\alpha = 2.08 \times 10^{-5}$)	48.87	94.09
	0.100	36.49% ($\alpha = 1.67 \times 10^{-4}$)	20.66	93.21
	0.500	17.65% ($\alpha = 2.08 \times 10^{-2}$)	53.00	56.48
	0.800	19.26% ($\alpha = 8.53 \times 10^{-2}$)	41.10	26.51
	1.000	25.31% ($\alpha = 1.67 \times 10^{-1}$)	41.40	20.87

lies in the compression of the Hilbert space on the network. To define the Hilbert space of the simplified quantum walk algorithm on complex networks, it is first necessary to determine the number of walking directions available for each node in the network. For the network $\tilde{G} = (V, E^{\mathrm{T}})$ defined in Section 4.1, this section introduces the idea of aggregating neighboring nodes into a single entity: consider the neighbors of node j as a whole and denote them as $w_{N(j)}$. For any node $j \in V$, there are only two possible walking directions: one is a path that stays at node j itself, and the other is a path that moves toward the neighbors of j (where the neighborhood of j is considered as a single entity). This aggregation concept can be illustrated using Figure 4.9a, where the dashed concentric circles represent the immediate neighbors of node j viewed as a whole. Thus, the dimension of the Hilbert space on network $\tilde{G}$ equals the number of nodes in the network multiplied by the number of walking directions available per node, which is $2N$.

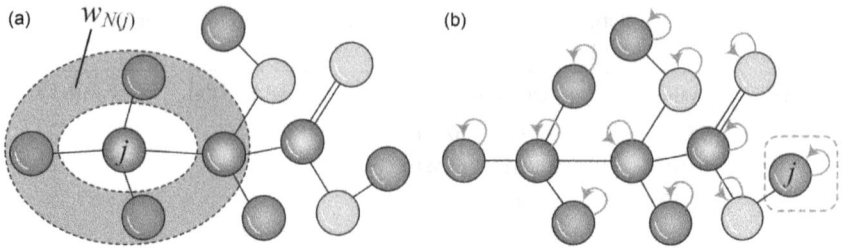

FIGURE 4.9 Design concept of the simplified quantum walk algorithm. (a) Aggregation design concept of the simplified quantum walk algorithm. (b) Network after adding self-loops to nodes [180].

When contrasted with the Hilbert space dimension of quantum walks on existing complex networks, the dimensionality of simplified quantum walks is significantly reduced. For instance, the QPageRank algorithm possesses a space dimension of N^2. Evidently, the simplified quantum walk algorithm achieves even lower dimensionality, especially when $\tilde{G}$ represents a dense network, where the advantage of low-dimensional space in the simplified quantum walk algorithm becomes more pronounced. In this network, each node has two possible directions of movement, indicated by $|0\rangle$ and $|1\rangle$ respectively. Suppose the initial state vector is an equal superposition, formally expressed as

$$|\psi(0)\rangle = \sum_{c=0}^{1} |c\rangle \otimes \sum_{j=1}^{N} \alpha_j(0)|j\rangle. \tag{4.23}$$

In this expression, $|j\rangle$ signifies the standard basis for node j; $\alpha_j(0)$ denotes the probability amplitude of node j at the initial moment. Since the initial state is an equal superposition, we have $\alpha_j(0) = 1/\sqrt{N}$.

The evolution operator for the simplified quantum walk consists of a coin operator and a shift operator. Referring to the definition form of the Grover algorithm, any element within the coin operator is defined as

$$G_{j,k}^{C} = -\delta_{j,k} + \frac{2}{|\Gamma(j,k)|}. \tag{4.24}$$

In Eq. (4.24), when $e(j,k) \notin E^{\mathrm{T}}$, then $\delta_{j,k} = 0$; otherwise, $\delta_{j,k} = 1$. Here, $\Gamma(j,k)$ represents the common neighbors between nodes j and k. According to the formulation in Eq. (4.24), the coin operator in the simplified quantum walk algorithm incorporates information on the similarity between nodes. As for the shift operator, suppose $|0\rangle$ and $|1\rangle$ indicate that the node either walks along a self-loop or toward its neighbor, respectively. Then the shift operator S is described as the summation over different walk directions:

$$S = \sum_{j=1}^{N} \left(|0,j\rangle\langle j,0| + \frac{1}{\sqrt{|N(j)|}} \sum_{k=1}^{N(j)} |1,j\rangle\langle k,1| \right). \qquad (4.25)$$

The shift operator calculated from Eq. (4.25) can be viewed as a large matrix composed of four block matrices of equal dimensions. These include two identity matrices located at the diagonal positions, and a matrix encoding the adjacency relationships within $\tilde{G}$; the other two blocks are zero matrices. The shift operator exhibits computational characteristics that offer distinct advantages. For instance, the identity matrices effectively add self-loops to each node in $\tilde{G}$, which can mitigate the negative impact of traceback. This principle is similar to the conclusions drawn in three-state quantum walks [97,99]. When the probability of a particle remaining at the current node increases, the probability of it staying at other nodes correspondingly decreases. For further details on traceback, refer to Section 2.2.3. Taking the network shown in Figure 4.9b as an example, node j acts as a peripheral node. After a single walk step, the particle can only jump from node j to its sole neighbor. Even so, after adding a self-loop to the peripheral node j, the probability of the particle jumping from node j to its neighbor is considerably reduced.

The evolution operator consists of the coin operator and the shift operator; thus, it can be defined as $U = S \cdot \left(G \otimes \hat{I} \right)$ after a finite number of walk steps. The measured probability on network $\tilde{G}$ can then be employed to score the similarity of each link within the set E^U. For any link $e(j,k) \in E^U$, the measurement result $P_{e(j,k)}$ signifies the probability of $e(j,k)$ being a missing link. Letting the evolution operator be applied t times, the similarity measurement result for link $e(j,k)$ is specified as

$$P_{e(j,k)} = \langle \hat{j} | U | \psi(t) \rangle = \langle \hat{j} | U^t | \psi(0) \rangle, \qquad (4.26)$$

$$|\hat{j}\rangle = \frac{|\Gamma(j,k)|}{|E^T|} \sum_{c=0}^{1} |j\rangle \otimes |c\rangle, \qquad (4.27)$$

where $\langle \hat{j} |$ is the conjugate transpose of $| \hat{j} \rangle$. It should be noted that, although Eq. (4.26) measures node j, this measurement result can effectively represent the score of link $e(j,k)$. This is because the neighborhood information of node j is already incorporated in Eq. (4.27). To minimize the negative impact of traceback, the step length of the simplified quantum walk is set to 2.

The link prediction performance of the simplified quantum walk algorithm proposed in this section is validated based on the AUC metric. Fourteen representative and innovative algorithms are selected as comparison methods, including Common Neighbors (CN) [181], Salton [182], Jaccard [183], Sorenson [184], Hub Promoted Index (HPI) [185], Hub Depressed Index (HDI) [186], Preferential Attachment (PA) [187], Adamic-Adar (AA) [188], Resource Allocation (RA) [186], Local Path (LP) [186], Katz [189], Average Commute Time (ACT) [190], Similarity by Cosine (Cos+) [191], and Neighbor Contribution (NC) [192]. The calculation methods for the similarity matrices of these algorithms are outlined in Table 4.3.

In Table 4.3, $N(j) \cap N(k)$ denotes the set of common neighbors between nodes j and k; for $z \in N(j) \cap N(k)$, z signifies a node within the common neighbor set of nodes j and k. In the Katz algorithm, υ is the weighting parameter for higher order paths in the network, typically set as the reciprocal of the largest eigenvalue of the adjacency matrix. In the ACT and Cos+ algorithms, l_{jk}^+ represents the element in the jth row and kth column of matrix L^+, where L^+ is the pseudoinverse of the Laplacian matrix of complex network G [168]. Additionally, in the NC algorithm, the symbol $\pi_{kj}^{(l)}$ indicates the probability that a particle, starting from node j, resides at node k within its l-hop neighborhood.

In the link prediction experiment conducted in this section, the values of the random variable μ are set to 0.9 and 0.8, indicating that 90% and 80% of the links in the link set E are used as the training set E^T, while the remaining 10% and 20% are utilized as the prediction set E^P. The experiment selects the Karate, Brain, Adjnoun, Email-univ, USAir, IceFire, Yeast, Email, and Economic networks as test datasets. The link prediction results based on the AUC metric are shown in Figures 4.10 and 4.11, where each color block represents the AUC value of a specific algorithm on a particular network. For ease of observation, the highest AUC value among the algorithms in each network dataset is marked with a red triangle. The higher the AUC value of an algorithm, the better its performance in predicting potential links in the network.

TABLE 4.3 Definitions of Similarity Matrices for Link Prediction

Algorithms	Calculation Formula	Introduction
CN	$\|N(j)\cap N(k)\|$	Number of common neighbors of a pair of nodes
Salton	$\dfrac{\|N(j)\cap N(k)}{\|N(j)\|\cdot\|N(j)\|}$	Local indicators based on common neighbors
Jaccard	$\dfrac{\|N(j)\cap N(k)\|}{\|N(j)\cup N(k)\|}$	Local indicators based on common neighbors
Sorenson	$\dfrac{2\|N(j)\cap N(k)\|}{\|N(j)\|+\|N(k)\|}$	Local indicators based on common neighbors
HPI	$\dfrac{\|N(j)\cap N(k)\|}{\min(\|N(j)\|,\|N(k)\|)}$	Local indicators based on common neighbors
HDI	$\dfrac{\|N(j)\cap N(k)\|}{\max(\|N(j)\|,\|N(k)\|)}$	Local indicators based on common neighbors
PA	$\|N(j)\|\cdot\|N(j)\|$	Preferred connection based on node degree value
AA	$\displaystyle\sum_{z\in N(j)\cap N(k)}\frac{1}{\log\|N(z)\|}$	Local algorithms based on common neighbors
RA	$\displaystyle\sum_{z\in N(j)\cap N(k)}\frac{1}{\|N(z)\|}$	Local algorithms based on common neighbors
LP	$A^2+\theta A^3$	Similarity algorithms based on local path
Katz	$(I-\upsilon A)^{-1}-I$	Similarity algorithms based on global path
ACT	$\dfrac{1}{l_{jj}^{+}+l_{kk}^{+}-2l_{jk}^{+}}$	Index based on global random walk
Cos+	$\dfrac{l_{jk}^{+}}{\sqrt{l_{jj}^{+}\cdot l_{kk}^{+}}}$	Index based on global random walk
NC	$\displaystyle\sum_{l=2}^{3}\left(\frac{\sum_{z\in N(j)}\frac{\|N(z)\|}{\max(N(j))}}{2M}\pi_{jk}^{(l)}+\frac{\sum_{z\in N(k)}\frac{\|N(z)\|}{\max(N(k))}}{2M}\pi_{kj}^{(l)}\right)$	Index based on local random walk

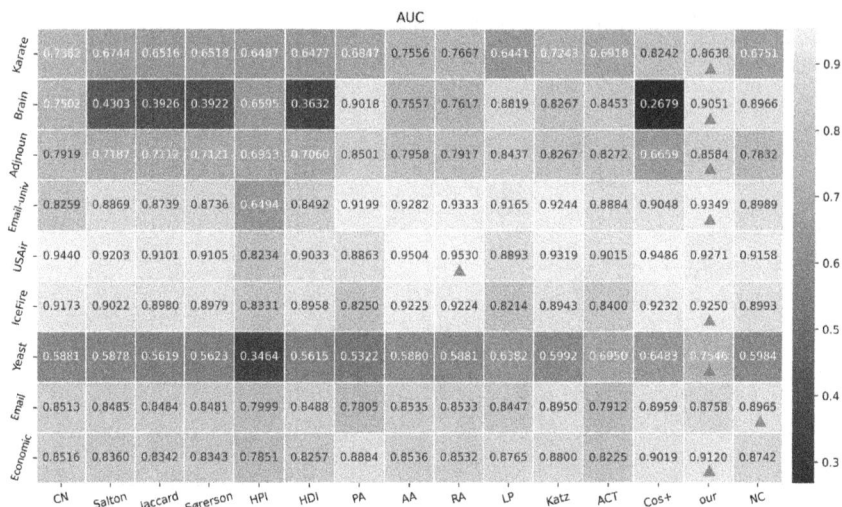

FIGURE 4.10   AUC results of various algorithms when $\mu = 0.9$ [180].

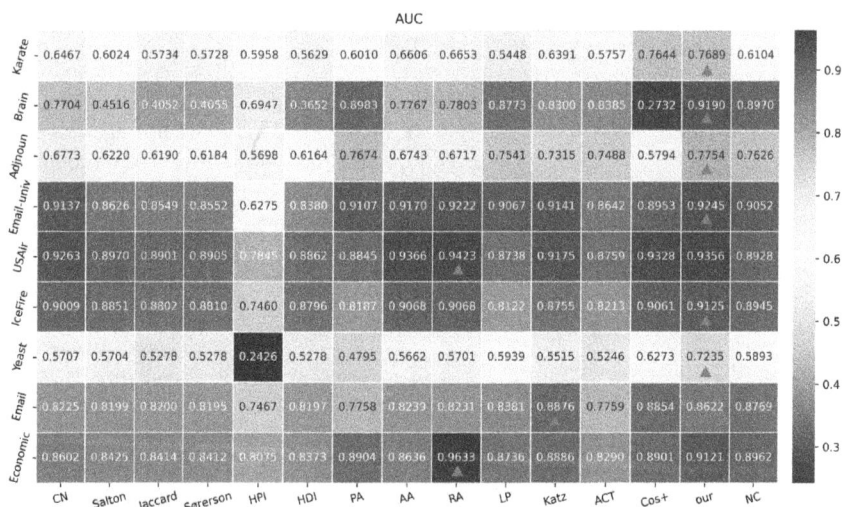

FIGURE 4.11   AUC results of various algorithms when $\mu = 0.8$ [180].

As illustrated in Figure 4.10, the simplified quantum walk algorithm suggested in this chapter achieves the highest AUC values across seven network datasets: Karate, Brain, Adjnoun, Email-univ, Economic, IceFire, and Yeast networks. Similarly, in Figure 4.11, the presented algorithm outperforms the other 14 comparison algorithms on six network datasets, specifically Karate, Brain, Adjnoun, Email-univ, IceFire, and Yeast networks.

Analyzing from the perspective of different types of link prediction algorithms, the following conclusions can be drawn.

1. In comparison to algorithms based on local neighbors, such as CN, Salton, Jaccard, Sorenson, HPI, and HDI, these algorithms and the simplified quantum walk algorithm outlined in this chapter both incorporate common neighbor information between node pairs. Nonetheless, experimental results using the AUC metric demonstrate that the simplified quantum walk model in this chapter achieves significantly higher accuracy in link prediction than local neighbor-based algorithms. It is noteworthy that local information algorithms also include the LP algorithm, which is based on local paths. As shown in Figures 4.10 and 4.11, the accuracy of the LP algorithm in link prediction is not comparable to the simplified quantum walk algorithm discussed here. Additionally, the LP algorithm adopts a step length of 3, incorporating information from three-hop neighborhoods, whereas the simplified quantum walk algorithm uses a step length of only 2.

2. Compared to global algorithms (Katz) and random walk algorithms (ACT and Cos+), the simplified quantum walk algorithm discussed in this section still achieves the highest AUC results overall. Only on the Email network does the Katz algorithm slightly outperform the introduced algorithm, as shown in Figure 4.11.

3. It is worth noting the comparison between the simplified quantum walk algorithm and the NC algorithm in terms of the AUC metric. The NC algorithm is a local random walk method; however, it only surpasses the simplified quantum walk algorithm in AUC value on the Email network, as presented in Figure 4.10.

4. Although the RA algorithm is considered a highly competitive classical method, its overall performance remains inferior to that of the simplified quantum walk algorithm outlined in this chapter. Using the AUC performance of the RA algorithm and the simplified quantum walk algorithm on the USAir network in Figure 4.11 as an example, their AUC values are highly similar, at 0.9423 and 0.9356, respectively. The experimental results in Figures 4.10 and 4.11 indicate that the simplified quantum walk algorithm can achieve high accuracy in network link prediction tasks.

4.4 SUMMARY AND DISCUSSION

This chapter introduces three algorithms for link extraction in complex networks, including the Hadamard walk algorithm for extracting critical links in static networks and key nodes in dynamic networks (Section 4.2), the continuous-time quantum walk algorithm for link prediction (Section 4.3.1), and the simplified quantum walk algorithm for predicting missing links in complex networks (Section 4.3.2). Additionally, the quantum walk-based critical link mining algorithm for complex networks also comprises the link centrality algorithm based on the Holevo quantity, as defined by Lockhart et al. [193]. This algorithm evaluates links by removing the links under examination from the network, defining the initial states, density operators, and measurement results of each link according to two scenarios: "including the link under consideration" and "excluding the link under consideration." Referring to Eq. (3.41) in Section 3.3.2 on QJSD, when the individual being evaluated is replaced with a link, the importance of a removed link in the entire network can be measured. Since the definitional format and primary conclusions of the Holevo quantity link centrality algorithm are analogous to those discussed in Section 3.3.2, further details will not be reiterated.

The three-quantum walk-based link extraction algorithms presented in this chapter lead to a common conclusion: the measurement results of quantum walk on complex networks can reflect the topological characteristics of the network while also predicting the evolutionary trends of the network, such as potential future links. This implies the possibility of establishing an inverse problem related to quantum walk on complex networks. Given the known initial probability amplitudes and several measurement results from the quantum walk, it becomes feasible to deduce the network structure, specifically the link relationships between nodes. Figure 4.12 provides two simple cases where the original network structure is inferred based on initial amplitudes and probability distributions. The vertical axis of the two-dimensional coordinates represents the measured probability distribution, while the horizontal axis suggests different nodes. For instance, in Figure 4.12a, which depicts a line of length 7, when the initial probability amplitude for discrete-time quantum walk on this line and the measurement results after the 1st to 3rd steps are known, the characteristic that the probability is zero at even positions when the step length is odd can be utilized (refer to the experimental conclusions in Section 2.1.1) to conclude that the shape has seven nodes in a linear arrangement.

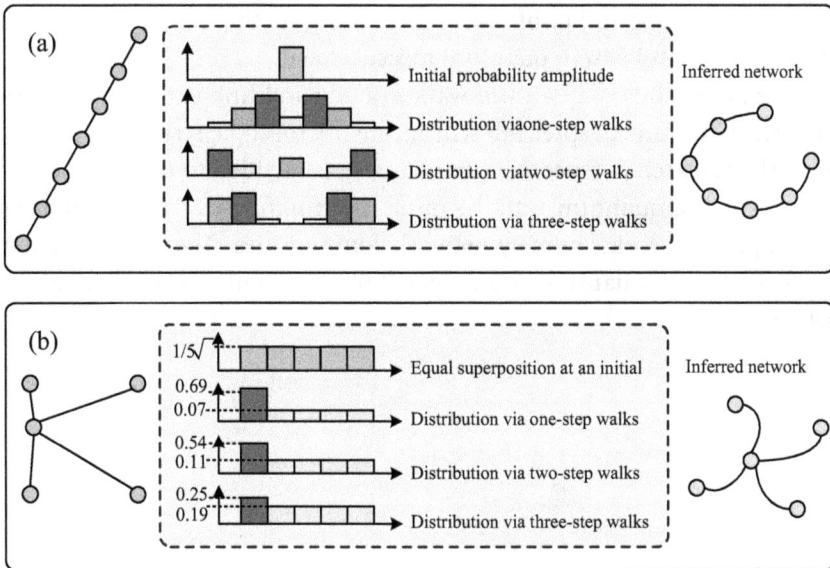

FIGURE 4.12 Illustration of network topology inferred from known distributions.

Similarly, for the star network depicted in Figure 4.12b, when the initial amplitude for continuous-time quantum walk is set as an equal superposition and the quantum measurement results for the first three moments are known, we can deduce that the original network is a star topology based on the higher probability of the central node compared to the edge (also known as hanging) nodes and the characteristic transfer of probability from the central node to the four edge nodes. If the known distribution of the quantum system is exposed during the process of generating random sequences through quantum walk, it might be possible to infer the original network topology based on this known distribution. This indicates that the generated keys are potentially vulnerable to being compromised. As highlighted in Section 2.2.1, although complex networks exhibit complexity, their substructures are often repetitive and similar. By summarizing the features of network structures and their corresponding distributions from a large amount of data, such features can be used as training samples. Machine learning methods could then, with a certain probability, ascertain the topology of complex networks based on known distributions. Consequently, the inverse problem of measurement results in quantum walk may pose a new challenge in quantum walk research, potentially sparking a notable influx of machine learning methods into this inverse measurement problem in quantum measurement.

The applications of quantum walk in node and link extraction within Chapter 3 and this chapter fall within the microscopic structural extraction. The research perspective can be further extended to the mesoscopic scale, utilizing quantum walk to mine meaningful subgraph structures in complex networks, namely network communities. Therefore, Chapter 5 will introduce quantum walk algorithms for community detection in networks.

Applications of Quantum Walks in Community Detection

5.1 COMMUNITY DETECTION: PROBLEM DESCRIPTION AND EVALUATION METRICS

Complex network community discovery, also known as community detection or network modularity mining, is a technique used to identify communities within complex networks. While the definition of a community remains debated [194], the characteristics of a community are clear: the nodes within the community are particularly closely connected, and the connections between the communities are very sparse. The subgraph in the dotted box in Figure 4.1a is a typical community structure that conforms to such characteristics. The saying "birds of a feather flock together" concisely summarizes the concept of communities. Nevertheless, real-life communities are highly complex. Taking the mysterious organization "The Frontiers of Science" from the science fiction novel *The Three-Body Problem* by Cixin Liu as an example, this organization was initially formed around a unified charter. Over time, however, members developed differing intentions and opinions, leading to internal divisions within "The Frontiers of Science" into factions like the Salvationists and the Survivors. There may also have been swing members who were aligned with multiple factions. This example from *The Three-Body Problem* illustrates that community structures have a temporal

DOI: 10.1201/9781003683902-7   **125** attribute and may undergo internal divisions or inter-community mergers over time. The swing members indicate that overlapping communities are possible, where a node can belong to multiple communities. The presence of leaders in the different factions of "The Frontiers of Science" suggests that, akin to the characteristics reflected at the macroscopic scale in complex networks, community structures exhibit repetitive, similar, and self-organizing features. The interpretation of *The Three-Body Problem* functions as a simple example of community structure characteristics. In reality, community discovery in networks is not limited to just overlapping communities, non-overlapping communities, and dynamic communities but also includes directed and undirected communities. Therefore, community discovery in practical applications involves several complex factors and is not merely an idealized map-like data clustering task. This makes the design and research of community discovery algorithms quite challenging.

The research goal of complex network community discovery is to divide N nodes in the complex network node set V into k independent non-empty subsets, $k \in \mathbb{Z}^+$, and each subset represents a community. Let C denote the community, and non-overlapping communities satisfy the following relationship: $C_1, C_2, \cdots C_k \subseteq V, C_1 \cap \cdots \cap C_k = \varnothing$, and $C_1 \cup \cdots \cup C_k = V$. The three quantum walk-based community discovery algorithms introduced in this chapter target non-overlapping communities. Current evaluation metrics for community discovery include internal and external evaluations. Internal evaluation is specified by the modularity function, which assesses whether the community detection result aligns with community characteristics. The external evaluations are established on the ground-truth community data of the network. Modular functions, defined by Newman [195], are also known as the Q function. In a complex network $G = (V, E)$, let σ_i be the index of the community to which a node belongs, and A represents the adjacency matrix of the network. The proportion of internal links within the community to the entire network is defined as

$$\frac{\sum_{j,k} A_{j,k} \delta(\sigma_j, \sigma_k)}{\sum_{j,k} A_{j,k}} = \frac{\sum_{i,j} A_{j,k} \delta(\sigma_j, \sigma_k)}{2M}. \tag{5.1}$$

In Eq. (5.1), if nodes j and k belong to the same community $\delta(\sigma_j, \sigma_k) = 1$, for any $j, k \in V$, otherwise $\delta(\sigma_j, \sigma_k) = 0$. When the community structure is fixed, in a complex network G with random links, the probability of a link existing between nodes j and k is denoted as $|N(j)||N(k)|/2M$. Thus, the modularity function [195] is described as

$$Q = \frac{1}{2M} \sum_{j,k} \left[\left(A_{j,k} - \frac{|N(j)||N(k)|}{2M} \right) \delta(\sigma_j, \sigma_k) \right]. \tag{5.2}$$

Empirical results show that the modularity function typically ranges between 0.3 and 0.7. The modularity function Q can not only be used to evaluate the accuracy of community detection results but also function as a heuristic function to achieve a community partition that maximizes modularity. In 2011, Lancichinetti et al. argued that maximizing modularity is not suitable as heuristic information for community discovery [196]. This is because, when a resolution parameter is introduced, the modularity function features the following issues: when the resolution parameter is small, the community discovery results tend to merge smaller communities; conversely, larger communities are more likely to be split into smaller ones. Hence, the evaluation of community discovery results often involves both internal and external metrics. The difference between the community division results of the algorithm and the real community data can be quantified by the difference in elements between the sets. Let the community division result of the algorithm be a distribution, where each element signifies the index of the community to which the current node belongs. At this point, the normalized mutual information (NMI) index is generally used to quantify the difference between the distribution and the real community division result. The NMI index is defined as

$$NMI(X, X') = \frac{2I(X, X')}{H(X) + H(X')}, \tag{5.3}$$

where the function $H(\cdot)$ represents the Shannon entropy of the community detection results; the calculation of $I(X, X')$ is given by $H(X) + H(X') - H(X, X')$, $NMI(X, X') \in [0,1]$. The larger the value of $NMI(X, X')$, the more accurate the community result X' is considered to be.

5.2 DISCRETE-TIME QUANTUM WALKS IN COMMUNITY DETECTION

This section will introduce two quantum walk algorithms for community detection. One is a two-stage quantum walk algorithm proposed by the authors of this book. The other is the Fourier quantum walk algorithm published in the journal *Physical Review Research*.

5.2.1 Two-Stage Quantum Walk Algorithm

The two stages of the two-stage quantum walk algorithm are the quantum walk without measurement defined by the authors of this book and k-means clustering. As the name suggests, quantum walks without measurement are quantum walks that do not involve measurement processes. As described in previous chapters on discrete-time quantum walks, the measurement result for any node in a quantum walk is a small numerical value. When quantum walks do not include measurement processes, each node in the network corresponds to a component in the evolution operator. If these vectors (components of the matrix) can carry similarity information between nodes, the quantum walk without measurement can act as input data for the k-means algorithm, and clustering vectors is precisely what the k-means algorithm excels at. Therefore, the main steps of the two-stage quantum walk algorithm for community detection in complex networks are: first, quantum walks without measurement are used to represent nodes as vectors, with similarity information between nodes serving as inspiration for this process; then, the clustering centers (centroids) of the k-means algorithm is determined based on the PageRank algorithm. Finally, the k-means algorithm is employed to cluster the vectors obtained in the previous processes, and the clustering results of the nodes are the communities identified by the algorithm. The flowchart of the two-stage quantum walk algorithm is displayed in Figure 5.1.

The definition of the two-stage quantum walk algorithm is as follows. In the first stage, the quantum walk without measurement defines the quantum state $|\psi_j\rangle$ of node j based on the connectivity of network G:

$$|\psi_j\rangle = \frac{1}{|N(j)|} \sum_{k \in N(j)} |k\rangle, \tag{5.4}$$

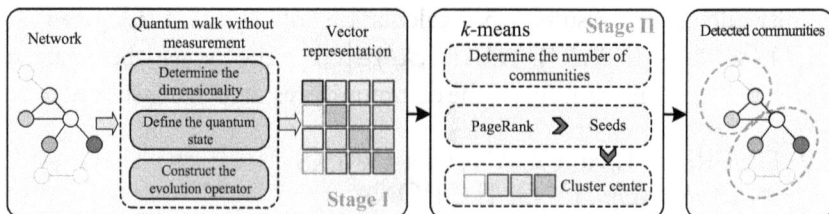

FIGURE 5.1 Flowchart of the two-stage quantum walk algorithm.

where $N(j)$ denotes the set of neighbor nodes of node j, $j \in V$, and $|k\rangle$ implies the standard basis relevant to node k, where node k is a neighbor of node j, that is, $k \in N(j)$. The result obtained from Eq. (5.4) is an N-dimensional vector, and the state that connects N nodes can be employed to construct a complete quantum state ψ. Let the function connecting N states be denoted as $\mathrm{conc}(\cdot)$; then, ψ is defined as

$$\psi = \sum_{j=1}^{N} \mathrm{conc}\left(|\psi_j\rangle\right), \tag{5.5}$$

where $\mathrm{conc}(\cdot)$ concatenates N quantum states $|\psi_j\rangle$ into an $N \times N$ matrix ψ, where $|\psi_j\rangle$ is a component of ψ. The final part of the quantum walk without measurement is the evolution process. To ensure that the vectors pertaining to the nodes contain similarity information, the common neighbor information between nodes is integrated during the evolution process. Let the adjacency matrix of network G be A, and the matrix representing the common neighbors between nodes be denoted as $A \times A = A^2$. The element in the j-th row and k-th column of the matrix A^2 is expressed as $A_{j,k}^2$, which implies the number of common neighbors between nodes j and k. As a result of this, following the form of the Grover operator in Eq. (1.15), the evolution operator for the quantum walk without measurement is given by

$$U = \frac{2}{A^2} \cdot \psi - \sum_{j=1}^{N} \mathrm{conc}\left[\left(\sum_{j=1}^{N} A_{j,*}^2\right) \cdot |\psi_j\rangle\right], \tag{5.6}$$

where $A_{j,*}^2$ represents the j-th row of the matrix A^2. The above content constitutes the definition of the first stage of the two-stage quantum walk without measurement. The second stage utilizes the PageRank algorithm to select the clustering centers for the k-means algorithm, and then the k-means algorithm is employed to complete the community detection task in the network. The method of overlapping community detection based on central node expansion has been proven to have the advantages of high speed and accuracy [197], so the second stage of the two-stage quantum walk algorithm adopts the PageRank algorithm to mine the central nodes of the network and use them as the clustering centers for the k-means algorithm.

The PageRank algorithm constructs the Google probability transition matrix based on random walks, and the calculation method of the Google probability transition matrix Gg is

$$Gg = \alpha E + \frac{(1-\alpha)}{N}\hat{i},$$ (5.7)

where α is the damping coefficient, typically set to 0.85; $\hat{i}$ is the identity matrix, and the construction method of matrix E is described as

$$E_{j,k} = \begin{cases} \frac{1}{N}, \sum_j A_{j,k} = 0 \\ \frac{A_{j,k}}{\sum_j A_{j,k}}, \sum_j A_{j,k} \neq 0 \end{cases}.$$ (5.8)

Drawing from this, the centrality of all nodes in the network is calculated using the Google probability transition matrix Gg, and the centrality scores of all nodes are sorted in descending order, with the high-score nodes being prioritized for selection as the clustering centers for the k-means algorithm. In complex networks, there often exists the rich-club phenomenon, where central nodes are densely connected to each other. If the distance between central nodes is too small, it can affect the clustering performance of the k-means algorithm. Therefore, a threshold $\wp$ should be set for high-centrality-score nodes to avoid clustering center aggregation. This requires that the PageRank measure values of any pair of centroid nodes satisfy $|PR(j) - PR(j)| > \wp$, for any nodes $j, k \in V$ and $j \neq k$, where $PR(j)$ represents the measure value of node j by the PageRank algorithm. When selecting k high-centrality-score nodes, i.e., specifying k clustering centers μ, each of the k clustering centers is an N-dimensional vector, namely, $\mu_1, \mu_2, ..., \mu_k \in \mathbb{R}^N$. The k-means algorithm is then employed to measure which clustering center each node j belongs to, with the calculation method being:

$$c(j) = \arg\min_{m, 1 \leq m \leq k} \|U_j - \mu_m\|^2.$$ (5.9)

When the Euclidean distance between the vector corresponding to node j and the clustering center μ_m is the shortest, it is determined that node j is part of the clustering center m, which is the community to which node j belongs.

In contrast to existing discrete-time quantum walks, the two-stage quantum walk algorithm exhibits the following advantages: (1) In the first stage of the algorithm, the Hilbert space dimension of the measurement-free quantum walk has been compressed to its minimum (N), making it easier to scale to larger networks compared to other discrete-time quantum walks. (2) The measurement process involves node-level computations. When the number of network nodes is extremely large, the computational overhead incurred by measurements becomes non-negligible. The measurement-free quantum walk not only eliminates the measurement process but also fully utilizes the evolution process to incorporate similarity information between nodes and represent them as vectors, providing valuable insights for accurately discovering community structures in complex networks. (3) The two-stage algorithm employs quantum walks as a data preprocessing step for the k-means algorithm, offering a novel approach to the design of existing quantum walk algorithms. This demonstrates that quantum walks can function as one of several steps within an algorithm, rather than relying entirely on quantum walks themselves and quantum devices to achieve target tasks.

5.2.2 Fourier Quantum Walk Algorithm

The Fourier quantum walk algorithm was proposed by Hatano's team from the University of Tokyo [111], in which the Fourier operator acts as the coin to drive the movement of particles across the nodes of a complex network. Hatano et al. also defined another quantum walk algorithm, which uses the Grover operator as the coin for quantum walks on complex networks. However, since the Fourier operator performs more excellently in degeneracy experiments and is more effective in the application of discovering communities in complex networks, this section only introduces the Fourier quantum walk algorithm. Regarding the degeneracy experiment, the method involves decomposing the evolution operator to obtain eigenvalues and projecting the real and imaginary parts of these eigenvalues onto a unit complex plane, finally observing their distribution on the unit circle. Similar experimental steps and results can be found in Section 2.1.1 of this book, which discusses the eigenvalue distribution.

According to the general framework in Section 2.3, the Hilbert space of the Fourier quantum walk algorithm is first defined based on the direct sum operation. This space is accumulated from the neighborhoods of each node:

$$\mathcal{H} = \oplus_{j=1}^{N} \mathcal{H}_j. \tag{5.10}$$

For an undirected complex network $G=(V,E)$, where $|E|=M$, the links connecting two nodes are non-directional. As such the total sum of the neighborhoods of all nodes equals twice the total number of network links, which indicates the Hilbert space dimension $D_{\mathcal{H}}=2M$. From this, the length of the quantum state in the Fourier quantum walk is $2M$. The Fourier quantum walk and the Hadamard walk similarly determine the space dimension using the direct sum operation. Thus, the initial quantum state and the probability amplitude settings of the Fourier quantum walk refer to the description in Eq. (4.6) from Chapter 4 on the Hadamard quantum walk and will not be repeated here.

Furthermore, the number of rows and columns of the evolution operator in the Fourier quantum walk is also $2M$. This evolution operator is obtained through matrix multiplication involving the coin operator C_F and the shift operator S. To ensure that the hopping process between node j and its neighboring node k is reversible, the shift operator S needs to act as a flip-flop between nodes j and k, satisfying $S|j,k\rangle=|k,j\rangle$. Since the Hilbert space of the Fourier quantum walk is constituted using the direct sum operation, the coin operator of the Fourier quantum walk is also formed using the direct sum operation on local coins. Therefore, the local coin operator in this quantum walk is defined as follows:

$$
C_j^F \begin{pmatrix} |j,k_1\rangle \\ |j,k_2\rangle \\ \vdots \\ |j,k_{d_j}\rangle \end{pmatrix} = \frac{1}{\sqrt{d_j}} \begin{pmatrix} 1 & 1 & \cdots & 1 \\ 1 & e^{\phi/d_j} & \cdots & e^{(d_j-1)\phi/d_j} \\ \vdots & \vdots & \ddots & \vdots \\ 1 & e^{(d_j-1)\phi/d_j} & \cdots & e^{(d_j-1)^2\phi/d_j} \end{pmatrix} \begin{pmatrix} |j,k_1\rangle \\ |j,k_2\rangle \\ \vdots \\ |j,k_{d_j}\rangle \end{pmatrix},
$$

$$(5.11)$$

where ϕ is the phase parameter; d_j is the number of neighboring nodes of node j. The global Fourier coin operator is constructed based on the local coin, similar to the form of Eq. (5.10), with the construction method being:

$$
C_F = \oplus_{j=1}^N \left(C_j^F \right). \tag{5.12}
$$

As a result, the evolution operator of the Fourier quantum walk is defined as

$$
U = S \cdot C_F. \tag{5.13}
$$

Finally, the measurement process of the Fourier quantum walk is designed. When a walker has taken t steps, the probability of the particle being located at node j is described as

$$P(j;t) = \sum_{k=1}^{N(j)} \left| \psi_{j,k}(t) \right|^2.$$ (5.14)

Derived from Eqs. (5.10) to (5.12), the components of each node in the quantum state are related to the number of its neighboring nodes. Hence, when measuring the probability of the particle being at node j as expressed in Eq. (5.14), it is necessary to sum the probability amplitudes of its neighboring nodes.

In the application of community discovery, the measurement result of the Fourier quantum walk is the mean value taken after the limit. The definers of the Fourier quantum walk believe that the measurement results are a superposition of behaviors at different oscillation frequencies, and such performance can be quantified by the eigenvalue decomposition result of the evolution operator after infinite steps, making the calculation of infinite average time meaningful [111]. Therefore, following Eq. (5.14), the average probability $\overline{p(i \rightarrow l)}$ of node i transitioning to node l after infinite evolution can be specified as

$$\overline{p(i \rightarrow l)} = \lim_{T \rightarrow \infty} \frac{1}{T} \frac{1}{k_i} \sum_{t=0}^{T-1} \sum_{m=1}^{k_l} \sum_{j=1}^{k_i} \left| \langle l \rightarrow m | U^t | i \rightarrow j \rangle \right|^2$$

$$= \frac{1}{k_i} \sum_{\mu=1}^{D} \sum_{m=1}^{k_l} \sum_{j=1}^{k_i} \left| \langle l \rightarrow m | \mu \rangle \right|^2 \left| \langle \mu | i \rightarrow j \rangle \right|^2.$$ (5.15)

In accordance with the construction forms of the variational quantum algorithm and the HHL quantum algorithm [3], the evolution operator can be decomposed into the composite of eigenstates. Thus, the evolution operator can be expressed as

$$U = \sum_{\mu=1}^{D_{\mathcal{H}}} | \mu \rangle e^{i\theta_\mu} \langle \mu |,$$ (5.16)

where $|\mu\rangle$ is the eigenvector, and $e^{i\theta_\mu}$ represents the eigenvalue of $|\mu\rangle$. Combining Eqs. (5.15) and (5.16), assuming that the measurement results of the Fourier quantum walk are a superposition of different oscillation frequencies, the result of taking the average after infinite evolution can be described as

$$\lim_{T\to\infty}\frac{1}{T}\sum_{t=0}^{T-1}e^{i(\theta_\mu-\theta_\nu)t}=\delta_{\mu\nu}. \qquad (5.17)$$

Equation (5.17) indicates that when the eigenvalues of the Fourier quantum walk evolution operator are distributed on the unit complex plane, these eigenvalues will not degenerate at positions −1 and +1. The above is the core assumption that the Fourier quantum walk can be used to discover community structures in networks.

5.2.3 Experiments and Analysis on Community Detection

This section employs the two-stage quantum walk and Fourier quantum walk algorithms to partition the well-known small-world network dataset, the Karate club, into communities. The Karate club split into two factions due to a dispute over whether to increase the fees, with each faction having a leader. Figure 5.2a shows the true community information of the Karate network, where nodes 1 and 33 are the leaders of the two factions, corresponding to the administrator and the president of the Karate club, respectively. Nodes 10 and 3, marked with arrows, are particularly difficult to classify by community discovery algorithms because node 10 has a link (friendship) in each faction, while node 3 has five links in each faction. Many topology-based unlabeled algorithms are prone to misclassifying these two nodes. Figure 5.2b presents the community discovery results of the Karate dataset using the two-stage quantum walk and Fourier quantum walk algorithms, which are consistent with the true community information and demonstrate excellent community discovery performance.

To further illustrate the advantages of the two-stage quantum walk algorithm outlined in this chapter and the Fourier quantum walk algorithm as proposed by Hatano et al. in community detection, a comparison was conducted using 15 algorithms through the NMI and modularity function Q. The selected 15 comparative algorithms are mainly divided into four categories: the first category includes quantum algorithms for community detection, such as quantum ant colony optimization (QACO) [198], quantum-behaved discrete multi-objective particle swarm

FIGURE 5.2   Community discovery results using the two-stage quantum walk and the Fourier quantum walk. (a) The ground-truth community of the Karate network. (b) Community division for the Karate network by two algorithms.

optimization (QDM-PSO) [199], and quantum genetic algorithm (QGA) [200]. The second category consists of community detection algorithms based on random walks, including the graph embedding with self-clustering (GEMSEC) method [201] and the WalkTrap method [202]. The third category comprises algorithms based on the modularity function, such as Louvain [203], greedy modularity [204], and Leiden [205]. The fourth category contains community detection algorithms based on information entropy, such as the significant community (SC) [206] and diffusion entropy reducer (DER) [207]. In addition to the above four categories, there are five other classical methods: spin-glass method [208], fluid method [209], Chinese Whispers method based on chaotic node clustering [210], the expectation-mixed (EM) model [211], and the spectral decomposition (SD) method [212]. Using the Karate Club and Football network (referred to as Football) data with real community information, the experimental results of the two-stage quantum walk and Fourier quantum walk

TABLE 5.1   Experimental Results of Complex Network Community Discovery Algorithm

Algorithms	Karate Dataset		Football Dataset	
	NMI	Q	NMI	Q
QACO	**1.000**	0.417	0.876	0.546
QDM-PSO	0.984	0.407	0.864	0.595
QGA	0.669	0.411	0.794	0.528
Two-stage quantum walk	**1.000**	0.371	0.516	0.603
Fourier quantum walk	**1.000**	0.371	—	—
GEMSEC	0.243	0.442	0.917	0.575
WalkTrap	0.826	0.339	0.898	0.602
Louvian	0.571	0.393	**0.982**	**0.604**
Greedy modularity	0.694	0.381	0.905	0.568
Leiden	0.579	0.407	0.969	0.603
SC	0.446	0.272	0.947	0.566
DER	**1.000**	0.371	0.414	0.349
Spin-glass	0.654	**0.420**	0.964	0.603
Chinese Whispers	0.836	0.319	0.938	0.568
Fuil	0.733	0.355	0.867	0.543
EM	**1.000**	0.371	0.841	0.509
SD	0.836	0.359	0.369	0.372

Note: The maximum values under different indicators in the table are marked in bold.

algorithms compared with the above 15 algorithms under the evaluation criteria of NMI and modularity function Q are provided in Table 5.1.

From the experimental results in Table 5.1, it can be observed that as quantum algorithms, QACO, the two-stage quantum walk, and Fourier quantum walk can accurately delineate communities in the Karate network. However, for the slightly more complex Football network, although QACO achieves the highest accuracy in community partitioning among quantum algorithms, the modularity of the two-stage quantum walk algorithm suggested in this chapter is the highest among these quantum algorithms, with a value of 0.603. It should be emphasized that while QACO fully and accurately partitions the communities in the Karate network, its modularity is higher than the modularity metric of the true communities, i.e., $Q > 0.371$. This is because the QACO algorithm is stochastic and requires multiple runs to average the results; thus, the community detection results obtained in each run are not identical, and the modularity metric of the true communities does not represent its maximum. In other words, although the computational results of community detection algorithms may differ from the real communities, they may uncover results with a very high degree of modularity, as evidenced by the performance of the two-stage quantum walk algorithm on the Football network and the spin-glass method on the Karate network. GEMSEC and WalkTrap, as random walk-based algorithms, perform better in community detection on the Football network than on the Karate network. Meanwhile, in modularity maximization algorithms, there are instances where the pursuit of modularity maximization fails to achieve high accuracy in partitioning network communities, such as the performances of Louvain, greedy modularity, and Leiden algorithms on the Karate network. Notably, Leiden, as an improved version of Louvain, still does not resolve this issue. This illustrates that modularity maximization is more suitable for describing whether the discovered communities exhibit the characteristics of "internally cohesive and externally sparse," rather than serving as the core design basis for community detection algorithms aiming to uncover networks with real community structures. Furthermore, when considering information-based community detection algorithms (SC, DER), it is evident that for networks with real communities, it is challenging to achieve a balance between the NMI and modularity metric results.

Additionally, in comparison with the Chinese Whispers, fluid method, and EM algorithm, the community division results of these three algorithms in the Football network outperform their performances in the Karate network, and all three execute worse than the two-stage quantum walk algorithm introduced in this chapter in the Karate network. The performance of the SD method in community detection is dictated by its characteristic of eigenvalue decomposition; when the number of communities is 2, it is easy to partition the associated nodes into two categories based on the positive and negative eigenvalues. Therefore, the SD method performs well in the Karate network. In contrast, the number of communities in the Football network is 11, which results in a significant decrease in the partitioning accuracy of the SD method.

5.3 CONTINUOUS-TIME QUANTUM WALKS IN COMMUNITY DETECTION

The continuous-time quantum walk designed by Faccin et al. [127], published in the *Physical Review X* journal in 2014, is currently the only known algorithm that employs continuous-time quantum walks for community detection in complex networks. Before introducing this algorithm, it is essential to understand the concept of "fluid" on complex networks. In simple terms, fluid is a change process with concepts of time and direction. For instance, information fluids on social networks record the cascade transmission of information, while traffic fluids on transportation networks carry information about vehicle passage. In the community detection algorithm based on continuous-time quantum walks, the authors of this book utilize changes in measurement probabilities, referred to as probability fluid, to capture community structures. Given the sparsity of connections between communities, the change in probability fluid between communities within a certain period is designed to be as stable as possible; while highly aggregated within communities, the fluid information changes faster within a period. Therefore, nodes within a community need to communicate with high fidelity. In summary, the core idea of the community detection algorithm based on continuous-time quantum walks is manifested in two aspects: (1) minimal variation in probability fluid outside the community; (2) maximum fidelity between nodes within the community. Faccin et al. designed two community detection algorithms based on continuous-time quantum walks, one optimizing for minimum probability fluid and the other maximizing fidelity [127].

5.3.1 Definition of Probability Fluid between Communities

Faccin et al. argued that the quantum community detection problem differs from its classical counterpart, seeking to find the Hilbert subspace corresponding to the node set of a community. The algorithm first views the complex network data as a closed quantum system and defines the Hamiltonian of this quantum system. When using the standard basis to represent nodes in the complex network, the Hamiltonian H is defined as

$$H = \sum_{i,j} H_{i,j} |i\rangle\langle j|, \tag{5.18}$$

where $H_{i,j}$ records the probability amplitude of a particle moving from node i to node j, $i \neq j$; when $i=j$, $H_{i,j}$ registers the energy value pertaining to the standard basis $|i\rangle$ of node i. Since the probability fluid captures the change in measurement probability over a period of time, the probability fluid T_A of any community A is defined as the difference between the measurement $p_A\{\rho(0)\}$ of the state vector at the initial time $(t=0)$ and the measurement result $p_A\{\rho(t)\}$ of the state vector at time t, namely,

$$T_\partial(t) = \frac{1}{2}\left| p_A\{\rho(t)\} - p_A\{\rho(0)\} \right|, \tag{5.19}$$

where the calculation method of the state vector $\rho(t)$ is

$$\rho(t) = e^{-iHt}\rho(0)e^{iHt}. \tag{5.20}$$

In Eq. (5.19), $p_A\{\rho(t)\}$ represents the measurement probability of community A at time t, and its calculation method is described as

$$p_A\{\rho\} = \text{tr}\{\Pi_A \rho\}, \tag{5.21}$$

where tr denotes the trace of a matrix, $\Pi_A = \sum_{i \in A} |i\rangle\langle i|$ and Π_A indicates the projection of the subspace of community A. When summing the probability fluids of all communities, the total probability fluid between communities can be defined as

$$T(t) = \sum_{A \in X} \frac{1}{2} \left| p_A\{\rho(t)\} - p_A\{\rho(0)\} \right|. \tag{5.22}$$

In addition, using the doubly stochastic transfer matrix $R(t)$ simplifies the formula for the probability fluid T_A of community A and yields a symmetric result $\tilde{R}_{i,j}(t)$. This process is implemented as

$$T_A(t) = \sum_{i \in A, j \notin A} \frac{R_{i,j}(t) + R_{j,i}(t)}{2} = \sum_{i \in A, j \notin A} \tilde{R}_{i,j}(t), \tag{5.23}$$

where $R_{i,j}(t) = \left| \langle i | e^{-iHt} | j \rangle \right|^2$. Inspired by the compactness formula of the transfer matrix in classical random walks [213], to keep the change in probability fluid between communities small, the objective function is formulated as

$$c_t^T(A, B) = \frac{T_A(t) + T_B(t) - T_{A \cup B}(t)}{|A||B|} = \frac{2}{|A||B|} \sum_{i \in A, j \in B} \tilde{R}_{i,j}(t). \tag{5.24}$$

When applying the long-time average method to process the results of Eq. (5.24), the doubly stochastic matrix can be integrated and expressed as

$$\hat{R}_{i,j}(t) = \frac{1}{t} \int_0^t R_{i,j}(t')dt'. \tag{5.25}$$

As t approaches infinity $(t \to \infty)$, $\hat{R}_{i,j}(t)$ can be described further as

$$\lim_{t \to \infty} \hat{R}_{i,j}(t) = \sum_k \left| \langle i | \Lambda_k | j \rangle \right|^2, \tag{5.26}$$

where Λ_k is the projection of the network Hamiltonian H onto the k-th eigenspace.

5.3.2 Definition of Fidelity within a Community

Another community discovery algorithm based on continuous-time quantum walks calculates community discovery results based on the intimacy formula of the fidelity of the quantum states of nodes within the community. Here, fidelity can be understood as a similarity score of the quantum states of nodes. The higher the similarity between nodes, the greater the likelihood that they belong to the same community. Over a given period, the total fidelity of all community quantum states is characterized as

$$F_X(t) = \sum_{A \in X} F_A(t) = \sum_{A \in X} F^2 \{ \Pi_A \rho(t) \Pi_A, \Pi_A \rho(0) \Pi_A \}, \tag{5.27}$$

where $\Pi_A \rho \Pi_A$ is the projection of the quantum state ρ onto the subspace associated with community A, and the calculation method for quantum state fidelity is given by the modulus of the inner product of two quantum states, expressed as

$$F\{\rho, \sigma\} = \operatorname{tr} \left\{ \sqrt{\sqrt{\rho} \sigma \sqrt{\rho}} \right\} \in \left[0, \sqrt{\operatorname{tr}\{\rho\} \operatorname{tr}\{\sigma\}} \right]. \tag{5.28}$$

Thus, based on the expression form of the compactness indicator, the objective function for maximizing community fidelity is described as

$$c_t^F(A, B) = \frac{F_{A \cup B}(t) - F_A(t) + F_B(t)}{|A||B|}$$

$$= \frac{2}{|A||B|} \sum_{i \in A, j \in B} \operatorname{Re} \left[\hat{\rho}_{i,j}(t) \rho_{j,i}(0) \right]. \tag{5.29}$$

where $c_t^F(A, B) \in [-1, 1]$, and

$$\hat{\rho}_{i,j}(t) = \frac{1}{t} \int_0^t dt' \rho_{i,j}(t'). \tag{5.30}$$

As t approaches infinity $(t \to \infty)$, $\hat{\rho}_{i,j}(t)$ can be expressed as

$$\lim_{t \to \infty} \hat{\rho}_{i,j}(t) = \sum_{k} \Lambda_k \rho_{i,j}(0) \Lambda_k. \tag{5.31}$$

Equations (5.24) and (5.29) optimize the community structure by integrating communities $\mathcal{A}$ and $\mathcal{B}$ to adjust the corresponding metric values.

To verify the effectiveness of the community discovery algorithm based on continuous-time quantum walks in this section, a random artificial network is first generated, and its community structure is shown in Figure 5.3a, which is divided into four non-overlapping communities. Derived from Eq. (5.3), by comparing the community information of the original random network, the NMI metrics of the continuous-time quantum walk algorithm in Figures 5.3b–d are 0.953, 0.82, and 0.85, respectively. It can be recognized that when the time parameter t takes a very small value, the continuous-time quantum walk algorithm, with the objective of minimizing the inter-community probability fluid, provides the highest partitioning accuracy. Specifically, Figure 5.3b presents the community discovery result when the probability fluid parameter $t \to 0$, where it is perceived that only one node is misclassified into another community. The experimental results in Figure 5.3c and d indicate that regardless of whether the probability fluid or fidelity method is used, setting the parameter t to a very large value $t \to \infty$ and averaging the measurement results will lead to a loss in community partitioning accuracy.

5.4 SUMMARY AND DISCUSSION

The community detection algorithm based on quantum walks introduced in this chapter once again demonstrates that quantum walks can effectively reflect the topological features of networks, highlighting the application value of quantum walks in complex network structure mining. However, considering the research work in Chapters 3–5, the applications of continuous-time quantum walks and discrete-time quantum walks in complex network structure mining should be distinguished. Broadly speaking, discrete-time quantum walks slightly outperform continuous-time quantum walks in applications on complex networks. By synthesizing the information propagation model based on continuous-time quantum walks from Section 3.3.3, the quantum link prediction algorithm from Section 4.3.1, and the community detection algorithm based on continuous-time quantum walks in this chapter, it is

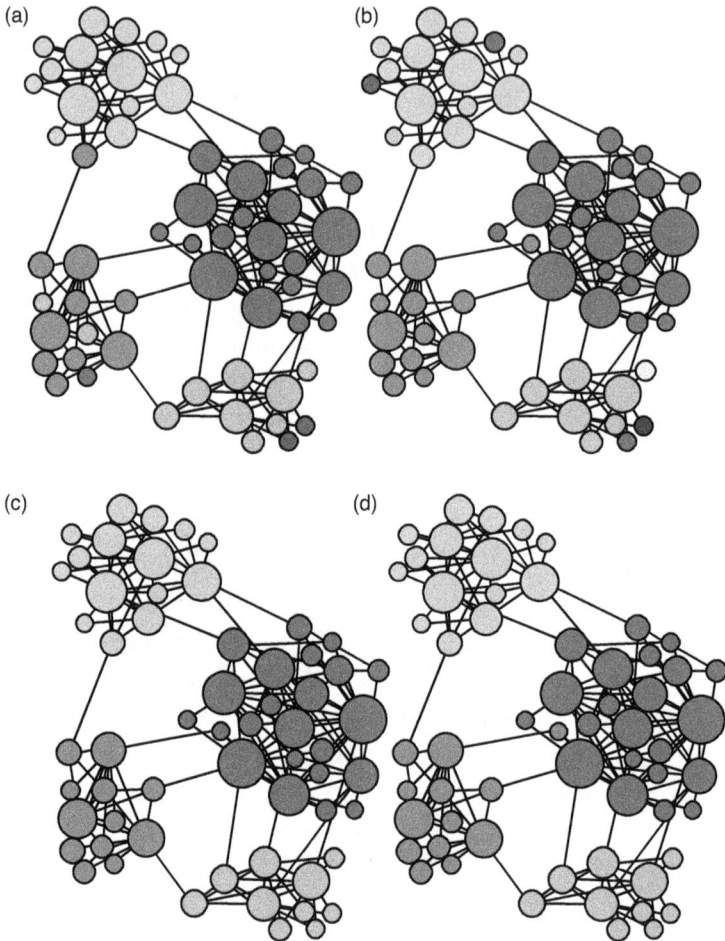

FIGURE 5.3 Original community and its community discovery result. (a) Random network. (b) Transport, $t \to 0$. (c) Transport, $t \to \infty$. (d) Fidelity, $t \to \infty$ [127].

evident that there are unresolved challenges for continuous-time quantum walks on complex networks. For example, the relationship between the parameter t in continuous-time quantum walks and the statistical characteristics of complex networks remains unclear. Researchers can only determine a suitable range for t based on metrics that favor evaluation indicators or directly define it as infinity without being able to infer this parameter's value range determined by network types. Furthermore, according to Euler's formula, the evolution results of continuous-time quantum walks belong to compound trigonometric functions with complex periodicity. This means that the measurement results of continuous-time quantum walks for nodes, links, and subgraph structures are oscillating. Therefore, in contrast to discrete-time quantum walks, continuous-time quantum walks are slightly weaker in applications on complex networks.

Brassard defined the Soufflé problem in quantum systems during their evolution in 1997, which describes the oscillation phenomena in measurement results of quantum circuits [214], where Soufflé is a type of baked egg-custard dish. Brassard believes that as the number of evolution steps increases, the probability amplitude for the target solution also augments. However, if the quantum system is measured too early, it leads to random collapse and sets the probability amplitude for the target solution to zero. The evolution and measurement processes of quantum systems are similar to the results of opening the oven too soon or overbaking a Soufflé, both of which cause the baking to fail. For quantum measurement, it fails to output the target solution effectively. Although the Soufflé problem targets the search efficiency of Grover search algorithm (detailed in Section 1.2.1), its description of the oscillation issue in continuous-time quantum walks on complex network structure mining still applies. This problem seems to be a fatal flaw in the application of continuous-time quantum walks on complex networks, but following the design ideas of quantum walk algorithms in this book, the oscillation issue in measurement results can be effectively avoided. Here, two design ideas are proposed for researchers to discuss: (1) Discretizing the parameters in continuous-time quantum walks, a technique clearly expressed in the heat diffusion model proposed by Ma et al. [215]. In the heat diffusion model, the network is also seen as a closed system. When only a few nodes are designated as initial nodes and assigned energy values, the energy transfers from higher to lower nodes along the links, a process where energy is conserved. Thus, the evolution process of the heat diffusion model is similar to that of continuous-time quantum walks, both represented by the product of the initial vector and the evolution operator to express the propagation result at a certain moment [215]. The formula's definition can be referred to as in Eq. (2.18). Existing cases show that the transmission results of the heat diffusion model can also mine structural information of complex networks [215–217]. The design idea of discretizing the parameter t in this model may become a crucial technique for the application of continuous-time quantum walks in complex networks. (2) According to the algorithm cases in Sections 3.3.3, 4.3.1, and 5.2.3, it is more beneficial for complex network structure mining when the parameter t in continuous-time quantum walks takes smaller values. The root cause is that continuous-time quantum walks rely on the adjacency matrix or Laplacian matrix of the network to complete evolution, belonging to a global network evolution (walk). The evolution process is similar to a breadth-first search of particles on the network; that is, when t takes smaller values, the probability amplitude can still be transmitted throughout the network and lead to traceback issues (see Section 2.2.3). In other words, under the condition that continuous-time quantum walks are global evolutions, adjusting the value of the parameter t can reduce the negative effects of traceback and improve the precision of continuous-time quantum walks in network structure mining tasks. In practical applications, the above two design ideas can be used in combination.

The research work in Chapters 3–5 collectively indicates that quantum walks effectively encompass complex network topological features during their evolution. This crucial information provides a solid foundation for the application of quantum walks in network representation learning and graph neural networks. Chapter 6 will introduce the application of quantum walks in role embedding, graph classification, and graph matching based on this premise.

Applications of Quantum Walks in Network Representation Learning

6.1 NETWORK REPRESENTATION LEARNING AND CLASSIFICATION TASKS

This chapter introduces network representation learning methods based on quantum walks. These methods form the foundation of research on feature extraction in networks using quantum walks and serve as a crucial basis for designing graph neural network (GNN) models and convolutional kernels with quantum walks. This section briefly introduces the fundamental concepts of traditional network representation learning and GNNs, as well as their interconnections.

6.1.1 Overview of Network Representation Learning and Graph Neural Networks

Network representation learning, also known as graph representation learning or network embedding, involves compressing graph structural information and using the compressed information to reconstruct the graph structure. In network representation learning, both compression and reconstruction are represented in vector or matrix form, and these processes are referred to as encoding and decoding, respectively.

Specifically, during the encoding phase, network representation learning encodes each node into a (low-dimensional) vector based on its

DOI: 10.1201/9781003683902-8

neighborhood topology, a process also known as node embedding. In the decoding phase, using the vectors obtained in the previous step, the similarity between any pair of embedded vectors is evaluated based on a given similarity matrix, aiming to determine if their similarity is reasonable. When the cumulative error between them is minimized, the node-embedding vectors can optimally reconstruct the network structure information and fulfill a specific training task. Established on this analysis, for any node $v \in V$ in a complex network $G = (V, E)$, the encoding process of network representation learning can be simply defined as a mapping of nodes into a d-dimensional vector space:

$$\text{enc} : V \to \mathbb{R}^d. \tag{6.1}$$

In the decoding phase, the similarity between a pair of nodes is primarily expressed through the product of their vector representations. This process can be described as

$$\text{dec} : \mathbb{R}^d \times \mathbb{R}^d \to \mathbb{R}^+. \tag{6.2}$$

GNNs are built upon network representation learning methods and can be seen as powerful models with tunable parameters for representing network structures. Introductory work on GNNs is extensive; for more details, refer to *Introduction to graph neural networks* [218]. To illustrate the relationship between network representation learning and GNNs, consider a simple yet foundational representation learning framework—shallow embedding. In shallow embedding, nodes are encoded as orthogonal vectors based on their identifiers, a process similar to the one-hot method, which is discussed in Chapter 1, Eq. (1.3), and its interpretation. When incorporating node feature information or local network structure information into the encoding phase to generate node embeddings, such frameworks form what is known as a GNN. In brief, GNNs integrate heuristic information beneficial for training tasks during the node-embedding phase, whereas shallow representation learning methods lack network topology or label information in their node embeddings.

6.1.2 Classification Tasks in Network Representation Learning and Graph Neural Networks

As mentioned in Section 1.2.3, quantum walks can be used to realize the Hamiltonian of a quantum system, which is a key element in quantum machine learning research [4–6]. This naturally led researchers to consider using quantum walks to design network representation learning methods and neural network models for extracting network feature information. In 2014, Schuld et al. designed a neural network model based on quantum walks [219], replacing traditional binary neurons with qubits and leveraging the coherent properties of quantum walks on graphs to accelerate information transfer within the neural network. However, this model is limited to low-dimensional regular graphs; when the dimension of the regular graph exceeds 7, the model cannot complete calculations in a reasonable time. Zhang et al. proposed a quantum-based subgraph convolutional neural network (QSCNN) using continuous-time quantum walks [220]. QSCNN considers both the global topology and local connectivity of the network, enabling high-accuracy node and network classification tasks. Nonetheless, QSCNN is only effective for specific families of tree-like graphs. These methods present new opportunities for applying quantum walks in machine learning, but their design relies entirely on the original quantum walk model. Due to constraints such as unitary transformations, these approaches lack flexibility in adapting measurement techniques, resulting in suboptimal performance on specific tasks.

This chapter introduces network representation learning and GNN methods based on quantum walks and analyze their performance in role embedding of nodes, network classification, and network isomorphism tasks. Since these three tasks involve network partitioning at either the micro-scale or mesoscale, they will collectively be referred to as network classification tasks in this section.

Role embedding, also known as role discovery or role detection [221], identifies structurally equivalent nodes that connect to other parts of the network in similar ways, classifying them under the same role. As an illustration, in a star-shaped graph, all peripheral nodes share one role, while the central node holds another. Role embedding differs from community detection [221], discussed in Chapter 5. For instance, in Figure 6.1a, nodes of the same color represent the same role; roles depend solely on structural similarity rather than connection density. In contrast, a community requires tightly connected nodes, as shown by the three dashed-line-enclosed node groups in Figure 6.1a, each forming a distinct community. Therefore, roles and communities are distinct concepts. The role embedding problem can be abstracted as mapping the N nodes in network G to a set $W = \{w_1, w_2, ..., w_\varepsilon\}$, where ε is the total number of roles, satisfying $\varepsilon \ll N$ and $\varepsilon \in \mathbb{Z}^+$. Let Φ denote the mapping

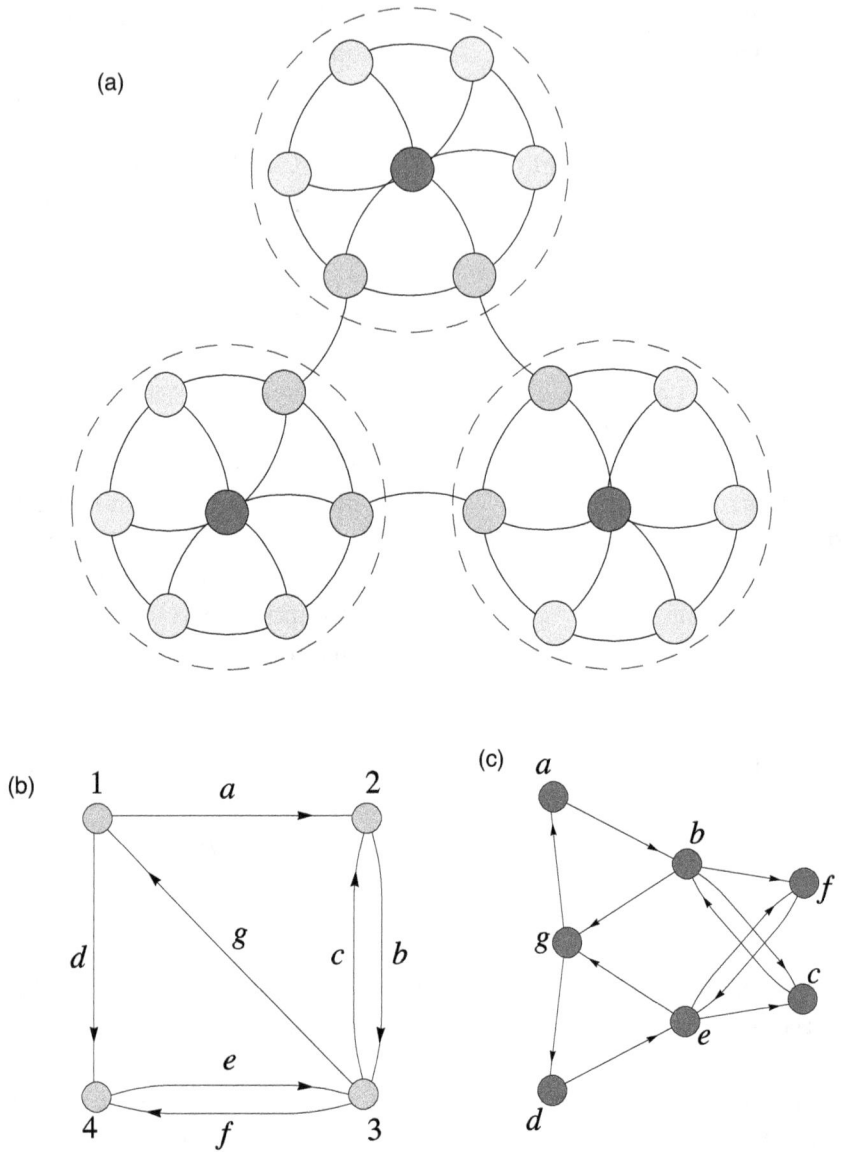

FIGURE 6.1 Illustrations of role embeddings and directed line graphs. (a) Illustration for the role embedding. (b) A directed graph. (c) A directed line graph.

function, and let X signify the matrix that represents structural features or attributes of the network. This process can be described as

$$\Phi: X \rightarrow W, \tag{6.3}$$

where X is an $N \times K$ matrix, and $K \leq N$. Let X_v denote a vector of length K for node v, where $v \in V$. If the role embedding task involves an attribute-free network, then K equals the total number of network nodes, that is, $K = N$. In Eq. (6.3), matrix X represents the embedding results for all nodes in the network, and W indicates the role detection outcomes for nodes based on the embedding information in X. Generally, the evaluation of role embeddings is built on the similarity measurement results between nodes [114], with common node similarity metrics provided in Table 4.3 in Chapter 4.

Network classification, also known as graph classification, aims to identify a mapping relationship between networks and their corresponding labels. Given a set of networks $\mathfrak{G} = \{G_i\}_{i=1}^n$ and corresponding labels $\mathcal{Y} = \{y_i\}_{i=1}^n$, each network is denoted as $G_i = (V, E, X)$, where the node set $|V| = N$ and the edge set $|E| = M$ define the network structure, and the feature information $X \in \mathbb{R}^{N \times d}$, with d representing the dimension of X. The task of network classification is to determine the classification label of the network by leveraging its adjacency matrix A and feature information X. In practical applications, graph networks often include binary and multi-class classifications. For binary classification, each network in $\mathfrak{G}$ is associated with a single classification label, represented as $\mathcal{Y} = \{y_{i,j}\}$ where $j = 1$. By extension, a network can also link to (m) classification labels, where $\mathcal{Y} = \{y_{i,j}\}$ and $j = 1, \cdots, m$.

Network isomorphism, also called graph isomorphism or graph matching, refers to a structural equivalence between two networks. Taking two graphs $G = (V, E)$ and $G' = (V', E')$ as examples, they are considered isomorphic if there exists a bijection $f: V \rightarrow V'$ such that for any $\forall v_1, v_2 \in V$, it implies $(v_1, v_2) \in E \Leftrightarrow (f_{v_1}, f_{v_2}) \in E'$. In addressing graph isomorphism tasks, the network to be analyzed can be transformed into a directed line graph to facilitate the construction of a high-dimensional feature space for the network [222,223]. A line graph functions as a secondary representation of the original graph, transforming edges into nodes and establishing link relationships based on the directional connections between edges. Figure 6.1b and c demonstrate examples of converting a directed graph into a directed line graph. For the line graph $G_L = (V_L, E_L)$ of a graph $G = (V, E)$, let the directed edge set of G be $E_d = \{e_d(u,v), e_d(v,u) | e(u,v) \in E\}$. Consequently, the node set V_L and edge set E_L of the line graph are defined as follows:

$$\begin{cases} V_L = E_d, V_L = E_d \\ E_L = \{(e_d(i,m), e_d(m,j)) \in E_d \times E_d\} \end{cases} \tag{6.4}$$

The network isomorphism problem has wide-ranging applications and has become a foundational topic for tasks such as network alignment and broad learning [216]. For instance, referring to the network described in Figure 4.2 of Chapter 4, a user can create independent accounts across different social media platforms. When the friendship (or follower) relationships among a group of users are identical on platforms like Sina Weibo and LinkedIn, the corresponding social networks on these platforms are considered isomorphic.

This chapter introduces quantum walk-based representation learning methods, neural network models, and graph convolutional kernels, focusing on the three-network classification tasks outlined above.

6.2 QUANTUM WALKS FOR NODE EMBEDDING

6.2.1 Quantum Walk-Based Node Similarity Estimation Algorithm

The node similarity estimation algorithm based on discrete-time quantum walks (QSIM), introduced in this section, was developed by a team from the National University of Defense Technology, China, and published in the journal *Applied Intelligence* in 2021 [113]. The QSIM algorithm is grounded in discrete-time quantum walks, using finite-step evolution results to characterize the similarity information between nodes. By comparing first-order and second-order node similarities, the QSIM algorithm demonstrates strong network reconstruction capabilities and excellent performance in estimating node similarities. The description of the QSIM algorithm follows the general framework presented in Section 2.3, with its four core components depicted in Figure 6.2.

In the QSIM algorithm, the Hilbert space is no longer composed of vectors but is instead defined as a matrix based on the dependency links between nodes. For a complex network G, its adjacency matrix A acts as the quantum state in the QSIM algorithm, represented as follows:

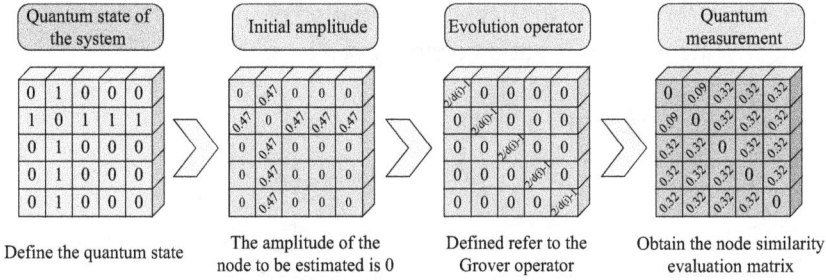

FIGURE 6.2   Core framework of the QSIM algorithm.

$$\psi_t = \sum_{|i,j\rangle} \alpha_{i,j}(t)|i,j\rangle, \tag{6.5}$$

where $|i,j\rangle$ contains the connectivity information of the adjacency matrix A, and $\alpha_{i,j}(t)$ represents the probability amplitude between nodes i and j at time t. In the QSIM algorithm, ψ_t is an $N \times N$ matrix. It includes a probability amplitude matrix φ_t, where φ_t records the probability amplitude $\alpha_{i,j}$ of the standard basis $|i,j\rangle$ after t steps of the walk, specifically given by

$$\varphi_t(i,j) = \alpha_{i,j}(t). \tag{6.6}$$

The QSIM algorithm specifies the node i to be evaluated by setting the initial probability amplitude. In this process, the initial probability amplitude of node i is set to 0, and the neighboring nodes of i equally share the remaining probability amplitude. The calculation method is described as follows:

$$\alpha_{i,j}(0) = \begin{cases} \dfrac{1}{\sqrt{N \cdot d(i)}}, & i \in V, \{i,j\} \in E \\ \\ 0, & \text{otherwise} \end{cases} \tag{6.7}$$

where $d(i)$ denotes the degree of node i. Equation (6.7) directly identifies the transmission path for the initial node probability amplitude, thereby eliminating the process of the probability amplitude returning from the neighboring nodes back to the initial node. This approach reduces the negative effects of traceback. Next, the evolution operator U for the QSIM algorithm is designed, which is modeled after the form of the Grover operator, incorporating the degree information of the nodes. The effect of the QSIM algorithm's evolution operator can be expressed as

$$|i,j\rangle \xrightarrow{U} \left(\frac{2}{d(i)} - 1 \right) |j,i\rangle + \frac{2}{d(i)} \sum_{\forall k \in N(i), k \neq i} |j,k\rangle, \tag{6.8}$$

where $\forall k \in N(i)$ indicates that any node k belongs to the neighborhood set $N(i)$ of node i. Further, one step of evolution in the QSIM algorithm is defined as $\psi_{t+1} = U\psi_t$.

The QSIM algorithm posits that, under the action of the Grover operator, it is essential to closely examine whether the measurement probability is higher for a particle returning to the initial node or to a neighboring node of the initial node after starting from the initial node. Through the initial probability amplitude setting in Eq. (6.7) and the evolution effect of Eq. (6.8), measurement results for the particle at different walk steps show a tendency to remain at the initial node. This approach in Eqs. (6.7) and (6.8) effectively reduces the negative impact of traceback, thereby enhancing the accuracy of the QSIM algorithm in calculating node similarity. This mechanism forms a crucial basis for the QSIM algorithm's capability to accurately characterize node similarity information.

Any quantum walk-based process for scoring network nodes relies on quantum measurement. In the QSIM algorithm, the measurement result of node i is stated as the summation of the probability amplitudes corresponding to node i and its neighborhood, defined as

$$p_t(i) = \sum_{j=1}^{N} |\alpha_{ij}(t)|^2$$

$$= \sum_{j=1}^{N} |\varphi_t(i,j)|^2. \tag{6.9}$$

The measurement result in Eq. (6.9) meets the requirement that the sum of probabilities equals 1, described as

$$\sum_{i=1}^{N} p_t(i) = \sum_{i=1}^{N} \sum_{j=1}^{N} |\varphi_t(i,j)|^2 = 1. \tag{6.10}$$

In conjunction with Eq. (6.9), let D denote the walk steps, given in the QSIM algorithm as the network's diameter. The similarity measure between any pair of nodes in the network, as determined by the QSIM algorithm, is expressed as

$$P(v,u) = \sum_{t=1}^{D} p_t(u). \tag{6.11}$$

Finally, the similarity score for any pair of nodes is recorded in the similarity matrix P using the following normalization method, with the elements on the diagonal set to 0:

$$P(v,u) = P(v,u) + \frac{1}{N-1} P(v,v), \forall u \in V, u \neq v. \tag{6.12}$$

Established on the above definitions, the QSIM algorithm represents a method for characterizing node feature information based on discrete-time quantum walks. This approach incorporates neighborhood topology information and leverages probability amplitudes to capture the local topological similarity between nodes. To validate the QSIM algorithm's capability in representing node similarity information, the node recommendation problem is introduced, using cosine similarity [191] as the evaluation criterion to assess the QSIM algorithm's performance in representing node similarity. The definition of cosine similarity is referenced from Table 4.3 in Chapter 4. For node recommendation, take a node v to be evaluated and identify similar nodes u, where the number of similar nodes u is flexible and non-unique. This experiment employs the Top-k most similar nodes as the benchmark. When the Top-k similar nodes set for a node v generated by a given algorithm closely matches the Top-k set determined by cosine similarity, the algorithm is considered to perform the recommendation task for node v with high accuracy. By performing this process for all nodes v in the network, the recommendation accuracy of the algorithm across all network nodes can be quantified [113]. This experiment compares the QSIM algorithm with Refex (recursive feature extraction) [224], Node2vec [225], and Role2vec [226]. The Dolphin, Polbooks, Football, Jazz, and Email networks are used as test datasets for the node recommendation experiments, with results based on first-order and second-order similarity provided in Tables 6.1 and 6.2, respectively.

TABLE 6.1   First-Order Similarity Calculation Results of the QSIM Algorithm [113]

Networks	Refex	Node2vec(1)	Node2vec(2)	Role2vec(1)	Role2vec(2)	QSIM
Dolphins	0.2527	0.0591	0.0645	0.4892	0.3925	**0.7849**
Polbooks	0.2024	0.1000	0.0833	0.4333	0.4595	**0.9429**
Football	0.3009	0.0974	0.0730	0.8017	0.8191	**1.0000**
Jazz	0.4416	0.1439	0.1328	0.5249	0.5032	**0.8820**
Email	0.0653	0.0072	0.0071	0.3864	0.2989	**0.7195**

Note: Bolded data indicates the maximum solution accuracy for the node recommendation task across different networks.

TABLE 6.2   Second-Order Similarity Calculation Results of the QSIM Algorithm [113]

Networks	Refex	Node2vec(1)	Node2vec(2)	Role2vec(1)	Role2vec(2)	QSIM
Dolphins	0.2132	0.0897	0.0833	**0.2952**	0.2845	0.2877
Polbooks	0.1849	0.0947	0.1154	0.3514	0.3312	**0.3587**
Football	0.2100	0.0932	0.0818	0.4792	0.4654	**0.5155**
Jazz	0.4084	0.1713	0.1641	0.4187	0.4139	**0.4974**
Email	0.0666	0.0118	0.0116	0.2047	0.1857	**0.2163**

Note: Bolded data indicates the maximum solution accuracy for the node recommendation task across different networks.

Founded on the experimental results presented in Tables 6.1 and 6.2, the QSIM algorithm achieves the highest average precision in node recommendation among the compared algorithms. Furthermore, examining the node recommendation results from Node2vec and Role2vec, it is evident that both algorithms exhibit similar capabilities in representing network topology when incorporating first-order and second-order information. Even with an increase in the walk depth (order) of Node2vec and Role2vec, neither algorithm is able to surpass the performance of the QSIM algorithm in the node recommendation task.

In fact, derived from Eqs. (6.5) and (6.8), the QSIM algorithm already incorporates the adjacency matrix of the complex network at the initial time, representing the global connectivity information of the network. The influence of first-order and second-order similarities on QSIM primarily manifests in the measurement results of one-step and two-step walks. Thus, this algorithm characterizes node similarity by integrating the topological features of the first two orders of neighboring nodes based on the global connectivity information of the network. Theoretically, this approach includes overlapping information from both local and global topologies, providing a potential advantage in representing the structural features of nodes. The performance of the QSIM algorithm, as shown in

Tables 6.1 and 6.2, indicates that quantum walks can effectively represent the structural characteristics of the local topology where nodes reside, embedding this information within the corresponding measurement results. This lays an important foundation for the further application of quantum walks in GNNs.

6.2.2 Quantum Walk-Based Role Embedding Algorithm

The definition of the role embedding via discrete-time quantum walk (RED) algorithm is similar to the concept of the QSIM algorithm in Section 6.2.1 and was likewise proposed by the team at the National University of Defense Technology [114]. The overall framework of the RED algorithm can be accessed in Figure 6.2, where the definitions of quantum state, probability amplitude, and evolution operator in the RED algorithm correspond to Eqs. (6.5)–(6.8) in Section 6.2.1 and are not reiterated here. A key difference between the RED and QSIM algorithms lies in the measurement phase. The RED algorithm posits that the evolution results of discrete-time quantum walks over complex networks can reflect a node's interaction information with the entire network and exhibit a consistency relationship between the node and the network as a whole. Since the measurement outcomes of discrete-time quantum walks do not converge, and averaging over limit steps is computationally expensive, the RED algorithm adopts a finite-step length d_g to ensure that the discrete-time quantum walk evolution satisfies quasi-periodicity.

When the RED algorithm evolves the initial quantum state of a discrete-time quantum walk for d_g steps, the probability of any node v after t steps is defined as the sum of the squared modulus of the probability amplitudes of node v and its neighborhood, that is,

$$p_v^t = \sum_j \left| \alpha_{vj}(t) \right|^2 = \sum_{j=1}^{N} \left| \varphi_t(v,j) \right|^2. \tag{6.13}$$

If the entire probability of node v from step 1 to step t is recorded in the distribution e_v^t, then the following distribution can be obtained:

$$e_v^t = \left[p_v^1, p_v^2, \cdots, p_v^{d_g} \right]. \tag{6.14}$$

After D steps of the walk, the average measurement probability of a given node represents the degree of similarity between this node and the initial node specified in Eq. (6.7):

$$c_v(i) = \sum_{t=1}^{D} p_i^t = \sum_{t=1}^{D} \sum_{j=1}^{N} |\varphi_t(i,j)|^2, \tag{6.15}$$

where the ith element of $c_v(i)$ records the similarity between node i and node v, where node i is a neighboring node of v; $i,v \in V$ and $i \in N(v)$, with $N(v)$ denoting the set of neighboring nodes of v. Based on Eq. (6.15), the RED algorithm defines the representation e_v^l of the local role of node v as follows:

$$e_v^l = \left[\max_1(c_v), \max_2(c_v), \cdots, \max_i(c_v), \max_{d_l}(c_v)\right], \tag{6.16}$$

where $\max_i(c_v)$ returns the ith maximum value of c_v, and $d_l = |e_v^l|$ implies the length of e_v^l. Finally, the RED algorithm combines e_v^g from Eq. (6.14) and e_v^l from Eq. (6.16) to generate the joint role embedding e_v of node v, calculated as follows:

$$e_v = f\left(e_v^g, e_v^l\right). \tag{6.17}$$

The essential difference between the RED and QSIM algorithms lies in the quantum measurement process. The quantum measurement process in the RED algorithm can be used to evaluate the role information of any node, based on the following: (1) using Eq. (6.7) to specify the node to be evaluated, which determines the starting position of the particle at the initial moment; (2) accumulating and averaging the measurement probabilities of the node across different walk steps. These two operations help to mitigate the traceback effect in quantum walks, leading the particle to tend to remain at the initially evaluated node during measurement, thereby accurately reflecting the structural characteristics of the node. To demonstrate the superior performance of the RED algorithm in role embedding, USAir, Polbooks, Adinoun, Jazz, and Email networks were selected as experimental datasets. The algorithms Node2vec [225], Rolx [227], Role2vec [226], and Graph Wave (GW) [228] were utilized as comparative methods. Using cosine similarity and inverse Euclidean distance as evaluation criteria, the role embedding results of the comparative algorithms and the RED algorithm are presented in Table 6.3.

As detailed in Table 6.3, the RED algorithm consistently achieves optimal performance based on both cosine similarity and inverse Euclidean distance. In contrast, the role embedding accuracy of the Node2vec and

TABLE 6.3   Role Embedding Results using Different Algorithms [114]

Networks	Evaluation Criteria	Node2vec	Rolx	Role2vec	GW	RED
USAir	Cosine similarity	0.002	0.697	0.053	0.514	**0.901**
	Inverse Euclidean distance	0.001	0.700	0.080	0.887	**0.909**
Polbooks	Cosine similarity	0.000	0.996	0.080	0.004	**1.000**
	Inverse Euclidean distance	0.000	0.996	0.086	0.994	**1.000**
Adjnoun	Cosine similarity	0.000	0.966	0.095	0.002	**0.966**
	Inverse Euclidean distance	0.000	0.969	0.103	0.979	**0.995**
Jazz	Cosine similarity	0.000	0.948	0.055	0.055	**0.979**
	Inverse Euclidean distance	0.000	0.948	0.057	0.977	**0.978**
Email	Cosine similarity	0.000	0.917	0.054	0.015	**0.983**
	Inverse Euclidean distance	0.000	0.916	0.080	0.978	**0.985**

Note: Bolded data indicates the maximum role embedding accuracy across different networks.

Role2vec algorithms is relatively low, while the performance of the Rolx and GW algorithms is highly unstable, with fluctuating accuracy. This suggests that the RED algorithm effectively identifies the topological structure information associated with each node and can perform role detection tasks with high accuracy under both cosine similarity and inverse Euclidean distance.

6.3 GRAPH NEURAL NETWORK AND GRAPH KERNEL BASED ON QUANTUM WALKS

6.3.1 Feature-Dependent Coin Quantum Walk Neural Network

The quantum walk neural network (QWNN), which relies on feature coins, encodes feature information of the nodes in the network topology within the quantum walk's coin. The evolution process of the quantum walk replaces the traditional information aggregation and message-passing processes in GNNs, thereby establishing a QWNN tailored for graph classification tasks [229]. This model, developed by Dernbach et al. [229] and published in *Applied Network Science* in 2019, is one of the most representative works in quantum walk-based GNNs. It uses discrete-time quantum walks as its foundation, with relevant theoretical details discussed in Sections 2.1.1 and 2.1.2. The QWNN model is a hybrid of quantum and classical computation and can be simplified into two main stages. In the first stage, the coin operator is used to learn node features, and in the second stage, the quantum walk integrates and propagates the neighborhood information of nodes.

In the first stage of the QWNN model, an independent coin operator is defined for each node, constructed based on the node's features, with the mapping relationship $f : X \rightarrow \mathbb{C}^{d \times d}$, where d denotes the maximum degree of nodes in the network. Taking node v_i as an example, where $\forall v_i \in V$, the coin operator pertaining to v_i is formulated as follows:

$$C_i = I - \frac{2f(v_i)f(v_i)^{\mathrm{T}}}{f(v_i)^{\mathrm{T}} f(v_i)}, \tag{6.18}$$

where I denotes the identity matrix. In the QWNN model, two different forms of expression are defined for the function $f(v_i)$. The first form uses betweenness centrality as heuristic information to design the parameter matrix W of this function, which is given by

$$f_1(v_i) = W^{\mathrm{T}} \mathrm{vec}\left(X_{\mathcal{N}(v_i)}\right) + b, \tag{6.19}$$

where $\mathcal{N}(v_i)$ is the column vector representing the feature connections of the neighborhood of node v_i. Each node has a unique, independent parameter matrix, and b is the bias vector for the function $f_1(\cdot)$. The second function, inspired by the similarity measure between node v_i and its neighborhood $\mathcal{N}(v_i)$, is described as

$$f_2(v_i) = X_{\mathcal{N}(i)} W X_i^{\mathrm{T}}, \tag{6.20}$$

where $W \in \mathbb{R}^{F \times F}$, and F signifies the number of features for each node; X is the matrix corresponding to the feature information.

In the second stage of the QWNN model, based on the coin operator constructed in the first stage and the fundamental theory of discrete-time quantum walks, a single walk step substitutes one training layer of the model. This process is achieved through alternating multiplications of the coin operator and the shift tensor. The encoding method of the shift tensor S for the graph structure is as follows: $S_{ujvi} = 1$ if and only if the ith node in the neighborhood of node v is u, and the jth node in the neighborhood of u is v; additionally, $S \in \mathbb{Z}_2^{N \times d \times N \times d}$. When T-step walk occurs, a superposition tensor ψ can be obtained, that is, $\psi = \left\{\psi^{(0)}, \psi^{(1)}, \ldots, \psi^{(T)}\right\}$. Each training layer uses the current superposition tensor $\psi^{(t)}$ as input. Using the coin operator set $C^{(t)}$ constructed in the first stage of the QWNN model, a single evolution of the quantum walk is identified as the product of the current superposition tensor, the coin operator set, and the shift tensor, expressed as

$$\psi^{(t+1)} = \psi^{(t)} C^{(t)} \cdot S. \tag{6.21}$$

The $\psi^{(t+1)}$ derived from Eq. (6.21) represents the output of the current training layer, which also functions as the input for the subsequent training layer.

The final component of the QWNN model is information transmission. Given the superposition tensor ψ, the propagation matrix is constructed by accumulating the superposition states. The calculation method is expressed below:

$$P = \sum_k \psi_{..k} \cdot \psi_{..k}, \tag{6.22}$$

where the calculation result P is a matrix, where each element P_{ij} represents the probability of remaining at node v_j starting from node v_i. Here, $\psi_{..k}$ denotes all components of the kth column in the matrix ψ. The propagation features are obtained through a composite calculation involving the matrix P and the feature vector X, with the calculation method as follows:

$$Y = h(PX + b), \tag{6.23}$$

where $h(\cdot)$ denotes a nonlinear function.

To evaluate the performance of the QWNN model in graph classification tasks, the ENZYMES, MUTAG, and NCI1 graph datasets were employed, with detailed descriptions of these datasets provided in the appendix of this book. The comparison methods selected include the graph convolutional network (GCN) [230], diffusion-convolutional neural network (DCNN) [231], graph attention network (GAT) [232], Weisfeiler–Lehman kernel (WLK) [233], and shortest-path kernel (SPK) [234]. The graph classification accuracy and standard deviation of the QWNN model and the above methods are provided in Table 6.4. In this table, QWNN(f_1) and QWNN(f_2) represent the classification results of the QWNN model when using Eqs. (6.19) and (6.20) as the node feature mapping functions, respectively.

Although the QWNN model's classification accuracy on the ENZYMES dataset is less satisfactory, it performs excellently on the MUTAG and NCI1 datasets. Particularly, when the function in Eq. (6.19) is used as heuristic information, the accuracy of the OWNN model on graph classification tasks shows a significant improvement. In summary, it can be concluded

TABLE 6.4   Graph Classification Accuracy and Standard Deviation of the QWNN Model and Comparative Algorithms [229]

Algorithms	ENZYMES Dataset	MUTAG Dataset	NCI1 Dataset
GCN	0.31±0.06	0.87±0.10	0.69±0.02
DCNN	0.27±0.08	0.89±0.10	0.69±0.01
GAT	0.32±0.04	0.89±0.06	0.66±0.03
WLK	**0.59 ± 0.01**	0.84±0.01	**0.85 ± 0.00**
SPK	0.41±0.02	0.87±0.01	0.73±0.00
QWNN(f_1)	0.26±0.03	**0.90 ± 0.09**	0.76±0.01
QWNN(f_2)	0.33±0.04	0.88±0.04	0.73±0.02

Note: Bolded values indicate the highest classification accuracy achieved across different datasets for graph classification tasks.

that this model effectively classifies the graph networks within the dataset. By integrating neighborhood information through particle walks on the graph, the QWNN model introduces a novel approach to quantum walk-based network feature extraction research. Furthermore, the model demonstrates that incorporating heuristic information beneficial to the target task can adjust the accuracy of the features extracted via quantum walks.

6.3.2 R-Convolution Kernel Based on Fast Quantum Walks

The fast quantum walk kernel (FQWK), based on rapid quantum walks, was published in 2022 in the journal *IEEE Transactions on Neural Networks and Learning Systems* [52]. It represents a significant new development in the application of quantum walks within GNNs and graph isomorphism neighborhoods. FQWK is a type of graph kernel based on discrete-time quantum walks; a graph kernel can be understood as a function that computes the inner product of graphs to measure their similarity. This section describes FQWK in three parts: representing the original graph network as a directed line graph, defining and implementing discrete-time quantum walks on the directed line graph, and using differences in the probability amplitudes of k-order neighborhood substructures to rapidly determine graph network isomorphism. For the basic concepts and transformation process of line graphs, readers may refer to Section 6.1.

As mentioned in Section 6.1, the line graph G_L provides a high-dimensional and dual representation of the original graph network G. Therefore, the number of rows and columns in the adjacency matrix corresponding to G_L equals the total number of links in G, that is, the number of nodes in G_L is $2|E|$, which represents the spatial dimension on the graph G_L. Quantum walks on the original network can be interpreted as encoding at the node level, whereas quantum walks on the directed line graph are equivalent to encoding at the link level. When a discrete-time quantum walk occurs on the directed line graph G_L of network G, let the standard basis for any link $e_d(u,v)$ in the line graph be denoted as $|u,v\rangle$; then, its quantum state is defined as

$$|\psi\rangle = \sum_{e_d(u,v)\in E_d} \alpha_{u,v}|u,v\rangle, \tag{6.24}$$

where $\alpha_{u,v}$ represents the complex-valued probability amplitude.

Referring to the Grover operator [2], the elements $U_{im,nj}$ in the evolution operator U for the discrete-time quantum walk can be formally expressed as follows:

$$U_{im,nj} = \begin{cases} A_{im}A_{nj}\left(\dfrac{2}{d_m} - \sigma_{ij}\right), & m = n \\ 0, & \text{otherwise} \end{cases}. \tag{6.25}$$

In this equation, $U_{im,nj}$ contains information about the probability amplitude transition from the directed link $e_d(i,m)$ to $e_d(n,j)$; here, d_m denotes the degree of node m. Regarding σ_{ij} in Eq. (6.25), it equals 1 only when $i = j$, and 0 otherwise. This defines the quantum walk component in the FQWK, which, as an R-convolution kernel, addresses a limitation of existing R-convolution kernels in recognizing the relative positions of subgraph structures. Specifically, as illustrated in Figure 6.3, the R-convolution kernel decomposes the original network into subgraphs when calculating graph structural similarity. In this example, three graph networks consist of two four-node closures and one three-node closure each, yet they are not isomorphic. Such cases are prone to misclassification in the original R-convolution kernel, as it cannot identify the relative positioning of subgraph structures.

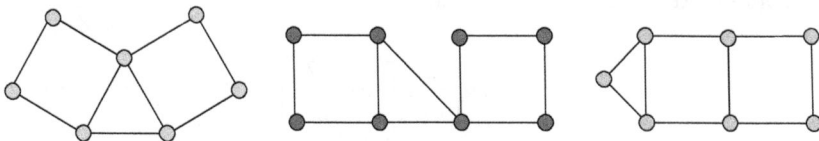

FIGURE 6.3   Non-isomorphic graphs with identical connected components.

In FQWK, the similarity of subgraph structures is primarily determined using information from k-level neighborhood-pair substructures. For any pair of nodes a and b in the line graph G_L, their k-level neighborhood-pair substructure, $S_{ab}^{(k)}$, consists of all paths of length k that begin at node a and end at node b. $S_{ab}^{(k)}$ is described as:

$$S_{ab}^{(k)} = \left\{ w \in W^{(k)} \middle| v_0 = a, v_k = b \right\}. \tag{6.26}$$

At this point, the probability amplitudes of all intermediate relay nodes along the path from node a to node b are summed. According to Eq. (6.25), the score between nodes a and b is defined as follows:

$$M_{ab} = \sum_{m=1}^{N} \sum_{n=1}^{N} U_{am,nb}. \tag{6.27}$$

When the walk step length (order) is specified, Eq. (6.27) can be further expressed as follows:

$$M_{ab}^{(t)} = \sum_{m=1}^{N} \sum_{n=1}^{N} U_{am;nb}^{t}, \tag{6.28}$$

where $M_{ab}^{(t)} \neq \left[M^t \right]_{ab}$. To achieve efficient computation of FQWK, the computational characteristics of matrix M at different orders are summarized below:

$$M^{(t)} = \begin{cases} ADA - 2D^{-1}, & t = 1 \\ MDA - A, & t = 2. \\ M^{(t-1)}DA - M^{(t-2)}, & t \geq 3 \end{cases} \tag{6.29}$$

In this equation, the mth element on the main diagonal of matrix D is equal to $2/d_m$, where d_m represents the degree of node m. Therefore, the diagonal matrix D is constructed as:

$$D = \text{diag}\left(\frac{2}{d_1}, \frac{2}{d_2}, \cdots, \frac{2}{d_N} \right). \tag{6.30}$$

In Section 2.3, the derived issues of quantum multi-step walks were analyzed, specifically the negative effects under superposed states. When the quantum walk steps $t \geq 3$, the evolution results of the system inevitably contain redundant information from first- and second-order subgraph structures. In Eq. (6.29), the computation for $t \geq 3$ effectively subtracts the redundant information of the lower order structural features, thus enhancing the structural characteristics within the line graph represented by the probability amplitude matrix M. When t takes different values in Eq. (6.29), using FQWK to address the graph isomorphism task allows the extraction of the t-order structural feature distribution of the graph network, which functions as input information for the FQWK function. FQWK expresses the R-convolution kernel function of the two graphs G_A and G_B to be determined as follows:

$$K_{FQWK}(G_A, G_B) = \sum_t K_t(G_A, G_B). \tag{6.31}$$

After t steps of the walk, a subgraph kernel function $K_t(G_A, G_B)$ is generated. This function is used to count all isomorphic neighborhood substructures with a walk steps of t. The computation method for this subgraph kernel function is outlined below:

$$K_t(G_A, G_B) = \sum_{m,n \in G_A} \sum_{u,v \in G_B} \Delta\left(S_{mn}^{(t)}, S_{uv}^{(t)}\right), \tag{6.32}$$

where $\Delta(\cdot)$ is a decision function that outputs 0 and 1, and its determination criteria are based on:

$$\Delta\left(S_{mn}^{(t)}, S_{uv}^{(t)}\right) = \begin{cases} 1, & M_{mn}^{(t)} = M_{uv}^{(t)} \\ 0, & \text{otherwise} \end{cases}. \tag{6.33}$$

In Eq. (6.33), $M_{mn}^{(t)}$ and $M_{uv}^{(t)}$ respectively denote the probability amplitudes pertaining to the neighborhood substructures $S_{mn}^{(t)}$ and $S_{uv}^{(t)}$ of order t. In other words, FQWK determines whether the two sets of neighborhood substructures are similar or isomorphic by comparing the differences in their corresponding probability amplitudes.

This section selects eight common or representative convolution kernel methods as comparative algorithms, including the random walk kernel (RWK), WLK [233], all graphlet kernel (AGK) [235], PATCHY-SAN convolutional neural network (PATCHY-SAN CNN, PSCNN) [236], deep graph convolutional neural network (DGCNN) [237], aligned subtree kernel (ASK) [238], quantum Jensen-Shannon kernel (QJSK) [152], and the edge-based matching kernel through discrete-time quantum walk (DQMK) [239]. Among these, ASK, QJSK, and DQMK are all quantum walk-based convolution kernels. The experiments in this section utilize the open-source networks Antiviral Screening Dataset (AIDS), Dihydrofolate Reductase Inhibitors Dataset (DHFR), Enzyme Protein Structure Dataset (ENZYMES), Mutagenicity Dataset (MUTAG), and Predictive Toxicology Challenge Male Mice Dataset (PTC_MM) as test datasets, with relevant descriptions provided in the appendix of this book. The graph classification accuracies of the aforementioned algorithms, along with FQWK, on these five networks are presented in Table 6.5.

According to the experimental results in Table 6.5, FQWK achieves higher graph classification accuracies on the AIDS, DHFR, ENZYMES, and PTC_MM datasets. When compared with other quantum walk-based convolution kernels (ASK, QJSK, and DQMK), although ASK outperforms FQWK on the MUTAG dataset, FQWK surpasses the other quantum walk-based convolution kernels in average graph classification accuracy across all five network datasets. Additionally, although PSCNN demonstrates comparable performance to FQWK on the AIDS dataset, its graph classification accuracies on the DHFR and PTC_MM datasets are significantly lower than those of other convolution kernel methods. Rooted in the above experimental analysis, FQWK, built upon directed line graphs, effectively identifies the local topological features of networks and accurately classifies the labels of graph networks.

TABLE 6.5   Average Graph Classification Accuracy of Different Algorithms [52]

Algorithms	AIDS	DHFR	ENZYMES	MUTAG	PTC_MM
RWK	80.00±0.28	78.20±0.61	14.20±0.42	80.56±0.72	61.59±0.86
WLK	98.89±0.70	79.39±0.57	37.69±0.62	83.22±0.89	61.72±0.81
AGK	99.07±0.07	78.20±0.61	28.88±0.61	82.01±0.90	63.65±0.82
PSCNN	**99.53±0.03**	77.66±0.14	15.50±0.09	83.16±0.11	59.41±0.34
DGCNN	98.50±0.03	78.28±0.46	40.12±0.11	77.78±0.51	54.55±0.76
ASK	96.74±0.12	78.17±0.61	30.26±0.60	**84.96±0.84**	61.15±0.81
QJSK	79.59±0.28	78.73±0.61	34.61±0.62	83.62±0.68	60.58±0.85
DQMK	79.99±0.67	78.15±0.64	28.91±0.71	76.42±0.88	61.09±0.73
FQWK	**99.53±0.06**	**80.87±0.57**	**41.55±0.61**	84.27±0.83	**63.77±0.78**

Note: Bolded data indicate the highest graph classification accuracies for each dataset.

6.4 SUMMARY AND OUTLOOK

From the perspective of expression and computation in network representation learning, the computational process based on quantum walk aligns with its related methods and offers potential training advantages. Network representation learning consists of two core stages: encoding and decoding. In the encoding stage, network nodes are represented as vectors, while in the decoding stage, the graph topology is reconstructed based on the encoded vectors to complete specific training tasks. The evolution of particles on the graph can be seen as the process of mapping nodes to a latent space, transforming them into embedding vectors. In this process, each network node is mapped to an embedding vector, and with the aid of coin operators and probability amplitudes as heuristic information, each embedding vector contains the local topological features of the corresponding node. The evolution result of particles on the graph is thus a matrix composed of embedding vectors for all nodes, which is consistent with the computational concept in the encoding stage of network representation learning. By utilizing the computational methods of continuous-time quantum walk, or the dimension-reducing discrete-time quantum walk, the length of each node's embedding vector can be constrained by the number of nodes in the network. In the decoding stage, spectral graph matrices or random walk methods can still be employed to evaluate node similarities and reconstruct the original graph data, thereby completing training tasks such as node classification, community detection, and personalized recommendations. As a result, the study of representation learning based on quantum walk in various tasks related to complex network structure extraction will emerge as a new research direction.

The research plan for network representation learning based on quantum walk can be further deepened in several ways. First, in the encoding phase, the encoded objects cannot only be nodes but also links, subgraphs, and community structures, which should be treated differently depending on the specific task. Second, by controlling the step length of the quantum walk or adopting multi-hop neighborhood integration methods from representation learning, the topological features of multi-hop neighborhoods can be incorporated into the embedding vectors. This approach would enrich the information contained in quantum walk-based representation learning methods. Moreover, during the encoding phase where the node and its neighborhood are integrated, partial sampling of the node's neighborhood instead of full sampling can be considered to improve the execution efficiency of quantum walk-based representation learning. Finally, treating the evolution operators and probability amplitude-related matrices and vectors as parameter arrays can help in two ways: adjusting the accuracy of the graph data reconstruction from embedding vectors and incorporating heuristic information from target tasks into the embedding vectors, thus enhancing the performance of quantum walk-based representation learning methods.

A substantial body of work in deep learning and GNNs suggests that if node-embedding vectors fail to accurately capture the structural information of the graph, any improvement in the training accuracy—regardless of the model's depth or robustness—will be severely limited. Apparently, the proper representation of nodes is the foundation and prerequisite for achieving high accuracy in training tasks. The network representation learning method based on quantum walk may offer new approaches and opportunities to address this challenge.

CHAPTER 7

Conclusions

THE CONVERGENCE OF QUANTUM walks and complex networks has led to new research ideas and solutions. Still, there are several issues that need to be addressed in applied research. This section presents five potential research hotspots and challenges for readers' reference, without repeating the conclusions, discussions, and prospects from previous chapters.

7.1 OPTIMIZATION OF QUANTUM WALK ON LARGE-SCALE COMPLEX NETWORKS

The appendix lists the complex network datasets involved in this book, and a considerable portion of these networks are sparse and small-scale. Nevertheless, some algorithms have already found it quite challenging to compute on these datasets. The parallelization of quantum walk computations will become an important research direction for its application in complex networks. Especially in complex problem environments, the computational resource consumption due to two-particle entangled states can be considerably reduced through parallel processing. However, there are few studies focusing on this area currently.

7.2 APPLICATION OF QUANTUM WALKS ON DIFFERENT TYPES OF COMPLEX NETWORKS

The types of complex networks involved in this book include bipartite networks, homogeneous networks, and dynamic networks. In reality, the types of complex networks are not limited to these and also include heterogeneous networks and multilayer networks, among others. It is still unclear

DOI: 10.1201/9781003683902-9

whether the existing quantum walk algorithms for complex networks can adapt to different types of networks. It is also uncertain whether the evolution of quantum walks on different types of complex networks can satisfy the unitary transformation conditions. Furthermore, some network data come with external information (such as weights, timestamps, and label information). There is no explicit design approach for how quantum walks can effectively include this information. Therefore, defining a generalized expression of quantum walks on different types of complex networks and enabling it to play an active role in network structure mining tasks is a challenging topic.

7.3 EXPLORING THE QUANTITATIVE RELATIONSHIP BETWEEN QUANTUM WALK MEASUREMENT RESULTS AND STATISTICAL CHARACTERISTICS OF COMPLEX NETWORKS

Numerous examples introduced in this book have demonstrated that the measurement results of quantum walks on complex networks can reflect the topological characteristics of the network. However, the relationship between the measurement results and the statistical properties of the network is still unclear. For example, it is difficult to determine which quantum walk measurement results on which type of network are more suitable for specific network structure mining tasks. Taking the application of Fourier quantum walks in community detection as an example [111], the phase parameter setting of the Fourier quantum walk plays a decisive role in the community division results, and the phase variation is also used in the synchronization control of complex networks to study the interaction relationship between an oscillator and a limited number of surrounding oscillators. Exploring and quantifying the relationship between measurement results and complex network statistical properties can further improve the computational accuracy of quantum walk algorithms in network structure mining tasks and promote their application in real life.

7.4 QUANTUM WALK ALGORITHMS FOR MINING COMBINED INDIVIDUALS IN COMPLEX NETWORKS

The quantum walk algorithms for complex network structure mining introduced in this book compute scores for nodes and links individually. However, when mining key nodes, the combination of the first and second most important nodes does not necessarily represent an important set of nodes, similar to how a team composed of the men's singles badminton

champion and the women's singles badminton champion cannot directly represent the mixed doubles badminton champion team. The current algorithms can only discover meaningful individual nodes or links but cannot detect meaningful combined individuals. Future quantum walk algorithms for mining combined individuals in complex networks will greatly promote research and development in this field.

7.5 DESIGN OF QUANTUM WALK ALGORITHMS ON COMPLEX NETWORKS REQUIRES "DUAL-TRACK PARALLELISM"

Since the generalized quantum walk algorithms are expected to run on future quantum devices, and conditions such as unitary evolution have somewhat limited the practical application and investigation of these algorithms, some quantum walk algorithms in this book do not strictly follow the unitary transformation conditions. Quantum walk algorithms that abandon unitary transformation conditions, though intuitively effective, are theoretically imperfect. Future algorithm designs based on quantum walks should focus on theoretical or practical applications, and these two design approaches should proceed in parallel, appropriately balancing such constraints. We should neither completely abandon unitary transformations for shortcuts nor rigidly adhere to constraints without flexibility. This dual-track design philosophy is also an important way for computer science researchers to quickly integrate into the field of quantum computing and leverage the advantages of quantum computing to accelerate the design and improvement of traditional algorithms.

The interesting and challenging research on the application of quantum walks in complex networks is not constrained to the five issues mentioned above. As research continues to develop and deepen, many new problems will emerge. Rather than considering this section as a conclusion, it might be more appropriate to view it as the beginning of research on the application of quantum walks in complex networks. We sincerely hope that readers from diverse fields will join the team of cross-research of quantum computing and network science, and work together on the long road and the boundless road.

Appendix

THE NETWORK DATASETS COVERED in this book can be divided into unlabeled and labeled categories, and this section provides a brief introduction to the statistical characteristics and basic information of the datasets covered in the book.

A.1 NETWORK DATASETS

1. Tribes Network: A network representing alliances between tribes.

2. Kangaroos Network: A social network depicting relationships within kangaroo groups.

3. Karate Network: A social network representing relationships among members of a karate club.

4. Chesapeake Network: A network of carbon exchanges between different biological species.

5. Dolphins Network: A social network of bottlenose dolphins.

6. Brain Network: A neural network of the rhesus monkey brain, where links represent fiber tracts connecting two neurons.

7. Polbooks Network: A network of purchase records for American political books, collected from Amazon during the 2004 U.S. presidential election.

8. Adjnoun Network: A network of co-occurrences between adjectives and nouns in the novel *David Copperfield*.

9. Football Network: A network of American football games between colleges in the year 2000.

TABLE A.1   Statistical Characteristics of Complex Network Datasets

Datasets	N	M	$\langle k \rangle$	d_{max}	d	c	ρ
Tribes	16	58	7.250	10	3	0.527	0.049
Kangaroos	17	91	10.706	15	3	0.841	−0.193
Karate	34	78	4.588	17	5	0.256	−0.475
Chesapeake	39	170	8.717	33	3	0.284	−0.375
Dolphins	62	159	5.129	17	8	0.259	−0.475
Brain	91	1,989	43.714	248	3	1.304	−0.343
Polbooks	105	441	8.400	25	7	0.488	−0.127
Adjnoun	112	425	7.589	49	5	0.173	−0.129
Football	115	613	10.661	12	4	0.407	−0.162
Enron-only	143	623	8.713	42	8	0.453	−0.019
Email-univ	167	3,251	38.934	139	5	0.685	−0.295
Jazz	198	2,742	27.697	100	6	0.618	0.020
Economic	259	2,942	19.560	108	6	0.395	0.018
USAir	332	2,126	12.807	139	6	0.749	−0.207
Infect-dublin	410	2,765	13.490	50	9	0.455	0.225
Metabolic	453	2,025	8.940	237	7	0.647	−0.225
Caltech36	769	16,656	43.320	248	6	0.409	−0.065
IceFire	796	2,823	81.982	122	9	0.209	−0.115
Email	1,133	5,451	9.622	71	8	0.220	0.078
Yeast	1,870	2,277	2.435	56	19	0.055	−0.161
Hamsterster	2,426	16,630	13.710	273	10	0.537	0.047
Facebook	2,888	2,981	2.064	769	9	0.001	−0.668
Wiki-vote	7,115	103,689	14.537	1,167	10	0.141	−0.083

Note: N denotes the number of nodes in the network, M indicates the number of links in the network, $\langle k \rangle$ characterizes the average degree of the network, d_{max} signifies the maximum degree in the network, d represents the diameter of the network, c quantifies the clustering coefficient of the network, and ρ measures the assortativity coefficient of the network.

10. Enron-only Network: A network of email interactions within the Enron corporation.

11. Email-univ Network: A network of email communications within a university team.

12. Jazz Network: A collaboration network of jazz musicians.

13. Economic Network: An economic network.

14. USAir Network: An air transportation network of the United States, where each node represents an airport and links represent flight routes between airports. The original network is weighted, with weights representing the frequency of flights.

15. Infect-dublin Network: A network of interactions among visitors at an art exhibition in Dublin, Ireland.

16. Metabolic Network: A metabolic network of the nematode *Caenorhabditis elegans.*

17. Caltech36 Network: A social network extracted from the Facebook platform.

18. IceFire Network: A fictional social network based on the novel *A Song of Ice and Fire.* Nodes represent words, and a link exists between two nodes if the corresponding words co-occur within 15 characters in the text.

19. Email Network: A network of email communications within a university.

20. Yeast Network: A protein interaction network of yeast, where nodes represent proteins and links represent metabolic interactions between proteins. The network contains 92 connected components, with the largest connected component including 2,375 nodes, which covers 90.75% of all nodes.

21. Hamsterster Network: A social network of friendships from the online social platform www.hamsterster.com, where nodes represent users and links represent friendships between users.

22. Facebook Network: The following relationships among users within the Facebook social network.

23. Wiki-vote Network: The user voting relationships generated during the Wiki administrator election process.

The statistical characteristics of the aforementioned network datasets are summarized in Table A.1. The unlabeled datasets are collected from the following sources.

1. Network Repository: http://networkrepository.com

2. Stanford University Social Networks Dataset: https://snap.stanford.edu/data/

3. KONECT: http://konect.cc/networks/

A.2 LABELED NETWORK DATASETS

The access links for the labeled network datasets are provided below, and their statistical characteristics are presented in Table A.2.

The TUDataset is available at: https://chrsmrrs.github.io/datasets/docs/datasets/

TABLE A.2   Statistical Characteristics of Labeled Network Datasets

Datasets	Number of Networks	Number of Classes	Average Number of Nodes	Average Number of Links	Data Description
AIDS	2,000	2	15.69	16.20	Biochemical small molecules
DHFR	756	2	42.43	44.54	Biochemical small molecules
ENZYMES	600	6	32.63	62.14	Bioinformatics data
MUTAG	188	2	17.93	19.79	Biochemical small molecules
NCI1	4,110	2	29.87	32.30	Biochemical small molecules
PTC_MM	336	2	13.97	14.32	Biochemical small molecules

References

1. Shor P W. Algorithms for quantum computation: Discrete logarithms and factoring. *Proceedings of the 35th Annual Symposium on Foundations of Computer Science*, 1994. IEEE, Santa Fe, USA.
2. Grover L K. A fast quantum mechanical algorithm for database search. *Proceedings of the 28th Annual ACM Symposium on Theory of Computing*, 1996. ACM, Philadelphia, USA.
3. Harrow A W, Hassidim A, Lloyd S. Quantum algorithm for linear systems of equations. *Physical Review Letters*, 2009, 103(15): 150502.
4. Schuld M, Sinayskiy I, Petruccione F. An introduction to quantum machine learning. *Contemporary Physics*, 2015, 56(2): 172–185.
5. Dunjko V, Briegel H J. Machine learning & artificial intelligence in the quantum domain: A review of recent progress. *Reports on Progress in Physics*, 2018, 81(7): 074001.
6. Biamonte J, Wittek P, Pancotti N, et al. Quantum machine learning. *Nature*, 2017, 549(7671): 195–202.
7. Nippon Yusen Kabushiki Kaisha Co., Ltd., Fujitsu Ltd. *Fujitsu and NYK Streamline Stowage Planning for Car Carriers by Leveraging Quantum-Inspired 'Digital Annealer'*. https://www.fujitsu.com/global/about/resources/news/press-releases/2021/0902-01.html, 2021.
8. Li W T, Huang Z G, Cao C S, et al. Toward practical quantum embedding simulation of realistic chemical systems on near-term quantum computers. *Chemical Science*, 2022, 13(31): 8953–8962.
9. Hefei Benyuan Quantum Computing Technology Co., Ltd. 合肥本源量子计算科技有限责任公司. A Detailed Interpretation of the Source Quantum Option Strategy Application 本源量子期权策略应用详细解读司. https://qcloud.originqc.com.cn/app/finTech, 2021.
10. Gong M, Wang S Y, Zha C, et al. Quantum walks on a programmable two-dimensional 62-qubit superconducting processor. *Science*, 2021, 372(6545): 948–952.
11. Chen J P, Zhang C, Liu Y, et al. Quantum key distribution over 658 km fiber with distributed vibration sensing. *Physical Review Letters*, 2022, 128(18): 180502.
12. Yan F, Venegas-Andraca S E. *Quantum Image Processing*. Berlin: Springer, 2020.

13. Yan F, Iliyasu A M, Venegas-Andraca S E. A survey of quantum image representations. *Quantum Information Processing*, 2016, 15(1): 1–35.

14. Li S Y 李士勇, Li Y 李研, Lin Y M 林永茂. *Intelligent Optimization Algorithms and Emergent Computation* 智能优化算法与涌现计算. Beijing 北京: Tsinghua University Press 清华大学出版社, 2019.

15. Xia P S 夏培肃. Quantum computing 量子计算. *Journal of Computer Research and Development* 计算机研究与发展, 2001, 38(10): 1153–1171.

16. Chandrashekar C M, Srikanth R, Laflamme R. Optimizing the discrete time quantum walk using a SU(2) coin. *Physical Review A*, 2008, 77(3): 032326.

17. Yan F, Li N, Hirota K. QHSL: A quantum hue, saturation, and lightness color model. *Information Sciences*, 2021, 577: 196–213.

18. Nielsen M A, Chuang I. *Quantum Computation and Quantum Information*. Cambridge: Cambridge University Press, 2010.

19. Williams C P, Clearwater S H. *Explorations in Quantum Computing*. Berlin: Springer, 1998.

20. Yan F 闫飞, Yang H M 杨华民, Jiang Z G 蒋振刚. *Quantum Image Processing and Application* 量子图像处理及其应用. Beijing 北京: Science Press 科学出版社, 2016.

21. Portugal R. *Quantum Walks and Search Algorithms*. Berlin: Springer, 2018.

22. Cerezo M, Arrasmith A, Babbush R, et al. Variational quantum algorithms. *Nature Reviews Physics*, 2021, 3(9): 625–644.

23. Rebentrost P, Mohseni M, Lloyd S. Quantum support vector machine for big data classification. *Physical Review Letters*, 2014, 113(13): 130503.

24. Bravo-Prieto C, García-Martí N D, Latorre J I. Quantum singular value decomposer. *Physical Review A*, 2020, 101(6): 062310.

25. Brassard G, Hoyer P, Mosca M, et al. Quantum amplitude amplification and estimation. *Contemporary Mathematics*, 2002, 305: 53–74.

26. Albash T, Lidar D A. Adiabatic quantum computation. *Reviews of Modern Physics*, 2018, 90(1): 015002.

27. Casalé B, Di Molfetta G, Kadri H, et al. Quantum bandits. *Quantum Machine Intelligence*, 2020, 2(1): 11.

28. Childs A M, Goldstone J. Spatial search by quantum walk. *Physical Review A*, 2004, 70(2): 022314.

29. Childs A M. Universal computation by quantum walk. *Physical Review Letters*, 2009, 102(18): 180501.

30. Buhrman H, Spalek R. Quantum verification of matrix products. arXiv preprint quant-ph/0409035, 2004.

31. Li M 李萌, Shang Y 尚云. Dynamics of coherence in quantum walk with two coins 两硬币量子游走模型中的相干动力学. *Journal of Computer Research and Development* 计算机研究与发展, 2021, 58(09): 1897–1905.

32. Wang Y N 王一诺, Song Z Y 宋昭阳, Ma Y L 马玉林, et al. Color image encryption algorithm based on DNA code and alternating quantum random walk 基于DNA编码与交替量子随机行走的彩色图像加密算法. *Acta Physica Sinica* 物理学报, 2021, 70: 32–41.

33. Wang H Q 王会权. *The Research on Controlling Probability Amplitudes of Quantum Walk—Model, Applications and Implementations* 量子漫步的概率幅调控技术研究—模型、应用和物理实现. Changsha 长沙: National University of Defense Technology 国防科技大学, 2016.

34. Aharonov Y, Davidovich L, Zagury N. Quantum random walks. *Physical Review A*, 1993, 48(2): 1687.

35. Shenvi N, Kempe J, Whaley K B. Quantum random-walk search algorithm. *Physical Review A*, 2003, 67(5): 052307.

36. Wang J B, Manouchehri K. *Physical Implementation of Quantum Walks*. Berlin: Springer, 2013.

37. Szegedy M. Quantum speed-up of Markov chain based algorithms. *Proceedings of the 45th Annual IEEE Symposium on Foundations of Computer Science*, 2004. IEEE, Rome, Italy.

38. Portugal R, Boettcher S, Falkner S. One-dimensional coinless quantum walks. *Physical Review A*, 2015, 91(5): 052319.

39. Douglas B L, Wang J B. Efficient quantum circuit implementation of quantum walks. *Physical Review A*, 2009, 79(5): 052335.

40. Berry D W, Childs A M. Black-box Hamiltonian simulation and unitary implementation. *Quantum Information & Computation*, 2012, 12(1/2): 0029–0062.

41. Duan B J, Yuan J B, Yu C H, et al. A survey on HHL algorithm: From theory to application in quantum machine learning. *Physics Letters A*, 2020, 384(24): 126595.

42. Wiebe N, Braun D, Lloyd S. Quantum algorithm for data fitting. *Physical Review Letters*, 2012, 109(5): 050505.

43. Lloyd S, Mohseni M, Rebentrost P. Quantum principal component analysis. *Nature Physics*, 2014, 10(9): 631–633.

44. De Bacco C, Larremore D B, Moore C. A physical model for efficient ranking in networks. *Science Advances*, 2018, 4(7): eaar8260.

45. Hefei Benyuan Quantum Computing Technology Co., Ltd. 合肥本源量子计算科技有限责任公司. Origin Quantum Cloud: Application Promotion Cloud 本源量子云: 应用推广云 https://qcloud.originqc.com.cn/app/finTech, 2020.

46. Deutsch D. Quantum theory, the Church–Turing principle and the universal quantum computer. *Proceedings of the Royal Society of London A Mathematical, Physical and Engineering Sciences*, 1985, 400(1818): 97–117.

47. Somma R D, Boixo S, Barnum H, et al. Quantum simulations of classical annealing processes. *Physical Review Letters*, 2008, 101(13): 130504.

48. Jordan P S. *Quantum Algorithm Zoo*. https://quantumalgorithmzoo.org, 2021.

49. Childs A M, Gosset D, Webb Z. Universal computation by multiparticle quantum walk. *Science*, 2013, 339(6121): 791–794.

50. Lovett N B, Cooper S, Everitt M, et al. Universal quantum computation using the discrete-time quantum walk. *Physical Review A*, 2010, 81(4): 042330.

51. Pearson K. The problem of the random walk. *Nature*, 1905, 72(294): 342–343.
52. Zhang Y, Wang L L, Wilson R C, et al. An R-convolution graph kernel based on fast discrete-time quantum walk. *IEEE Transactions on Neural Networks and Learning Systems*, 2022, 33(1): 292–303.
53. Abd El-Latif A A, Abd-El-Atty B, Venegas-Andraca S E, et al. Providing end-to-end security using quantum walks in IoT networks. *IEEE Access*, 2020, 8: 92687–92696.
54. Yang Y G, Pan Q X, Sun S J, et al. Novel image encryption based on quantum walks. *Scientific Reports*, 2015, 5(1): 7784.
55. Yang Y G, Xu P, Yang R, et al. Quantum Hash function and its application to privacy amplification in quantum key distribution, pseudo-random number generation and image encryption. *Scientific Reports*, 2016, 6(1): 19788.
56. Abd El-Latif A A, Abd-El-Atty B, Venegas-Andraca S E. A novel image steganography technique based on quantum substitution boxes. *Optics & Laser Technology*, 2019, 116: 92–102.
57. Abd-El-Atty B, El-Latif A, Ahmed A, et al. An encryption protocol for NEQR images based on one-particle quantum walks on a circle. *Quantum Information Processing*, 2019, 18: 272
58. Wang Y, Shang Y, Xue P. Generalized teleportation by quantum walks. *Quantum Information Processing*, 2017, 16: 221.
59. Shang Y, Wang Y, Li M, et al. Quantum communication protocols by quantum walks with two coins. *EPL (Europhysics Letters)*, 2019, 124(6): 60009.
60. Barnum H, Crepeau C, Gottesman D, et al. Authentication of quantum messages. *Proceedings of the 43rd Annual IEEE Symposium on Foundations of Computer Science*, 2002. IEEE, Vancouver, Canada.
61. Meijer H, Akl S. Digital signature schemes for computer communication networks. *ACM SIGCOMM Computer Communication Review*, 1981, 11(4): 37–41.
62. Feng Y Y, Shi R H, Shi J J, et al. Arbitrated quantum signature scheme with quantum walk-based teleportation. *Quantum Information Processing*, 2019, 18: 154.
63. Feng Y Y 冯艳艳, Shi R H 施荣华, Shi J J 石金晶, et al. Arbitrated quantum signature scheme based on quantum walks 基于量子游走的仲裁量子签名方案. *Acta Physica Sinica* 物理学报, 2019, 68: 68–76.
64. Li H J, Li J, Xiang N, et al. A new kind of universal and flexible quantum information splitting scheme with multi-coin quantum walks. *Quantum Information Processing*, 2019, 18(10): 316.
65. Krovi H, Magniez F, Ozols M, et al. Quantum walks can find a marked element on any graph. *Algorithmica*, 2016, 74(2): 851–907.
66. Ambainis A, Kempe J, Rivosh A. Coins make quantum walks faster. arXiv preprint quant-ph/0402107, 2004.
67. Magniez F, Nayak A, Roland J, et al. Search via quantum walk. *SIAM Journal on Computing*, 2011, 40(1): 142–164.
68. Magniez F, Nayak A, Richter P C, et al. On the hitting times of quantum versus random walks. *Algorithmica*, 2012, 63(1): 91–116.

69. Buhrman H, Dü R R C, Heiligman M, et al. Quantum algorithms for element distinctness. *SIAM Journal on Computing*, 2005, 34(6): 1324–1330.

70. Ambainis A. Quantum walk algorithm for element distinctness. *SIAM Journal on Computing*, 2007, 37(1): 210–239.

71. Magniez F, Santha M, Szegedy M. Quantum algorithms for the triangle problem. *SIAM Journal on Computing*, 2007, 37(2): 413–424.

72. Wong T G. Faster quantum walk search on a weighted graph. *Physical Review A*, 2015, 92(3): 032320.

73. Rhodes M L, Wong T G. Quantum walk search on the complete bipartite graph. *Physical Review A*, 2019, 99(3): 032301.

74. Rapoza J, Wong T G. Search by lackadaisical quantum walk with symmetry breaking. *Physical Review A*, 2021, 104(6): 062211.

75. Chakraborty S, Novo L, Ambainis A, et al. Spatial search by quantum walk is optimal for almost all graphs. *Physical Review Letters*, 2016, 116(10): 100501.

76. Tanaka H, Sabri M, Portugal R. Spatial search on Johnson graphs by continuous-time quantum walk. *Quantum Information Processing*, 2022, 21(2): 74.

77. Moradi M, Annabestani M. Möbius quantum walk. *Journal of Physics A: Mathematical and Theoretical*, 2017, 50(50): 505302.

78. Li P C 李鹏程. *Continuous-Time Quantum Walks on Complex Networks* 复杂网络上的连续时间量子游走. Shanghai 上海: Fudan University 复旦大学, 2013.

79. Berry S D, Wang J B. Quantum-walk-based search and centrality. *Physical Review A*, 2010, 82(4): 042333.

80. Venegas-Andraca S E. Quantum walks: A comprehensive review. *Quantum Information Processing*, 2012, 11(5): 1015–1106.

81. Li D 李丹. *Analysis and Applications of Discrete-Time Quantum Walks* 离散型量子漫步模型分析及其应用. Beijing 北京: Beijing University of Posts and Telecommunications 北京邮电大学, 2016.

82. Abd El-Latif A A, Abd-El-Atty B, Mazurczyk W, et al. Secure data encryption based on quantum walks for 5G Internet of Things scenario. *IEEE Transactions on Network and Service Management*, 2020, 17(1): 118–131.

83. Yang Y G, Bi J L, Chen X B, et al. Simple hash function using discrete-time quantum walks. *Quantum Information Processing*, 2018, 17: 189.

84. Wong T G, Tarrataca L, Nahimov N. Laplacian versus adjacency matrix in quantum walk search. *Quantum Information Processing*, 2016, 15(10): 4029–4048.

85. Omar Y, Paunković N, Sheridan L, et al. Quantum walk on a line with two entangled particles. *Physical Review A*, 2006, 74(4): 042304.

86. Štefaná K M, Kiss T, Jex I, et al. The meeting problem in the quantum walk. *Journal of Physics A: Mathematical and General*, 2006, 39(48): 14965.

87. Bisio A, Dariano G M, Mosco N, et al. Solutions of a two-particle interacting quantum walk. *Entropy*, 2018, 20(6): 435.

88. Lahini Y, Verbin M, Huber S D, et al. Quantum walk of two interacting bosons. *Physical Review A*, 2012, 86(1): 011603.

89. Chandrashekar C M, Busch T. Quantum walk on distinguishable non-interacting many-particles and indistinguishable two-particle. *Quantum Information Processing*, 2012, 11(5): 1287–1299.

90. Rodriguez J P, Li Z J, Wang J B. Discord and entanglement of two-particle quantum walk on cycle graphs. *Quantum Information Processing*, 2015, 14(1): 119–133.

91. Sun X Y, Wang Q H, Li Z J. Interacting two-particle discrete-time quantum walk with percolation. *International Journal of Theoretical Physics*, 2018, 57(8): 2485–2495.

92. Costa P C S, De Melo F, Portugal R. Multiparticle quantum walk with a gaslike interaction. *Physical Review A*, 2019, 100(4): 042320.

93. Carson G R, Loke T, Wang J B. Entanglement dynamics of two-particle quantum walks. *Quantum Information Processing*, 2015, 14(9): 3193–3210.

94. Berry S D, Wang J B. Two-particle quantum walks: Entanglement and graph isomorphism testing. *Physical Review A*, 2011, 83(4): 042317.

95. Li D, Zhang J, Guo F Z, et al. Discrete-time interacting quantum walks and quantum Hash schemes. *Quantum Information Processing*, 2013, 12(3): 1501–1513.

96. Li D, Zhang J, Ma X W, et al. Analysis of the two-particle controlled interacting quantum walks. *Quantum Information Processing*, 2013, 12(6): 2167–2176.

97. Falkner S, Boettcher S. Weak limit of the three-state quantum walk on the line. *Physical Review A*, 2014, 90(1): 012307.

98. Inui N, Konno N. Localization of multi-state quantum walk in one dimension. *Physica A: Statistical Mechanics and its Applications*, 2005, 353: 133–144.

99. Inui N, Konno N, Segawa E. One-dimensional three-state quantum walk. *Physical Review E*, 2005, 72(5): 056112.

100. Machida T, Chandrashekar C M. Localization and limit laws of a three-state alternate quantum walk on a two-dimensional lattice. *Physical Review A*, 2015, 92(6): 062307.

101. Zeng M, Yong E H. Discrete-time quantum walk with phase disorder: Localization and entanglement entropy. *Scientific Reports*, 2017, 7(1): 12024.

102. He Z M, Huang Z M, Li L Z, et al. Coherence evolution in two-dimensional quantum walk on lattice. *International Journal of Quantum Information*, 2018, 16(02): 1850011.

103. Attal S, Petruccione F, Sinayskiy I. Open quantum walks on graphs. *Physics Letters A*, 2012, 376(18): 1545–1548.

104. Hatano N, Obuse H. Delocalization of a non-Hermitian quantum walk on random media in one dimension. *Annals of Physics*, 2021, 435: 168615.

105. Saha A, Mandal S B, Saha D, et al. One-dimensional lazy quantum walk in ternary system. *IEEE Transactions on Quantum Engineering*, 2021, 2: 1–12.

106. Mochizuki K, Bessho T, Sato M, et al. Topological quantum walk with discrete time-glide symmetry. *Physical Review B*, 2020, 102(3): 035418.

107. Ribeiro P, Milman P, Mosseri R. Aperiodic quantum random walks. *Physical Review Letters*, 2004, 93(19): 190503.

108. Kendon V. Decoherence in quantum walks–A review. *Mathematical Structures in Computer Science*, 2007, 17(6): 1169–1220.
109. Konno N. Quantum walks. *Sugaku Expositions*, 2020, 33(2): 135–158.
110. Kendon V. Quantum walks on general graphs. *International Journal of Quantum Information*, 2006, 4(5): 791–805.
111. Mukai K, Hatano N. Discrete-time quantum walk on complex networks for community detection. *Physical Review Research*, 2020, 2(2): 023378.
112. Schofield C, Wang J B, Li Y Y. Quantum walk inspired algorithm for graph similarity and isomorphism. *Quantum Information Processing*, 2020, 19(9): 281.
113. Wang X, Lu K, Zhang Y, et al. QSIM: A novel approach to node proximity estimation based on discrete-time quantum walk. *Applied Intelligence*, 2021, 51(4): 2574–2588.
114. Wang X, Jian S L, Lu K, et al. RED: Learning the role embedding in networks via discrete-time quantum walk. *Applied Intelligence*, 2021, 52(2): 1493–1507.
115. Liu K, Zhang Y, Lu K, et al. MapEff: An effective graph isomorphism agorithm based on the discrete-time quantum walk. *Entropy*, 2019, 21(6): 569.
116. Chawla P, Mangal R, Chandrashekar C M. Discrete-time quantum walk algorithm for ranking nodes on a network. *Quantum Information Processing*, 2020, 19(5): 158.
117. Wong T G. Equivalence of Szegedy's and coined quantum walks. *Quantum Information Processing*, 2017, 16(9): 215.
118. Wong T G, Santos R A M. Exceptional quantum walk search on the cycle. *Quantum Information Processing*, 2017, 16(6): 154.
119. Paparo G D, Müller M, Comellas F, et al. Quantum Google in a complex network. *Scientific Reports*, 2013, 3(1): 2773.
120. Paparo G D, Müller M, Comellas F, et al. Quantum Google algorithm: Construction and application to complex networks. *The European Physical Journal Plus*, 2014, 129(7): 150.
121. Loke T, Tang J W, Rodriguez J, et al. Comparing classical and quantum PageRanks. *Quantum Information Processing*, 2017, 16(1): 25.
122. Bai X M 白晓梅. *Scholarly Impact Evaluation and Prediction Based on Social Network Analysis* 基于社会网络分析的学术影响力评估与预测. Dalian 大连: Dalian University of Technology 大连理工大学, 2017.
123. Xu X P, Liu F. Continuous-time quantum walks on Erdös–Rényi networks. *Physics Letters A*, 2008, 372(45): 6727–6732.
124. Faccin M, Johnson T, Biamonte J, et al. Degree distribution in quantum walks on complex networks. *Physical Review X*, 2013, 3(4): 041007.
125. Razzoli L, Paris M G, Bordone P. Transport efficiency of continuous-time quantum walks on graphs. *Entropy*, 2021, 23(1): 85.
126. Matrasulov D, Stanley H E. *Nonlinear Phenomena in Complex Systems: From Nano to Macro Scale*. Berlin: Springer, 2014.
127. Faccin M, Migdał P, Johnson T H, et al. Community detection in quantum complex networks. *Physical Review X*, 2014, 4(4): 041012.

128. Chakraborty S, Novo L, Roland J. Finding a marked node on any graph via continuous-time quantum walks. *Physical Review A*, 2020, 102(2): 022227.

129. Osada T, Coutinho B, Omar Y, et al. Continuous-time quantum-walk spatial search on the Bollobás scale-free network. *Physical Review A*, 2020, 101(2): 022310.

130. Li X, Chen H W, Ruan Y, et al. Continuous-time quantum walks on strongly regular graphs with loops and its application to spatial search for multiple marked vertices. *Quantum Information Processing*, 2019, 18(6): 195.

131. Sánchez-Burillo E, Duch J, Gómez-Gardenes J, et al. Quantum navigation and ranking in complex networks. *Scientific Reports*, 2012, 2(1): 605.

132. Izaac J A, Zhan X, Bian Z H, et al. Centrality measure based on continuous-time quantum walks and experimental realization. *Physical Review A*, 2017, 95(3): 032318.

133. Liu H, Xu X H, Lu J A, et al. Optimizing pinning control of complex dynamical networks based on spectral properties of grounded Laplacian matrices. *IEEE Transactions on Systems, Man, and Cybernetics: Systems*, 2021, 51(2): 786–796.

134. Chen W, Lakshmanan L V S, Castillo C. *Information and Influence Propagation in Social Networks*. Berlin: Springer, 2013.

135. Chen C Y 陈超洋, Zhou Y 周勇, Chi M 池明, et al. Review of large power grid vulnerability based on complex network theory 基于复杂网络理论的大电网脆弱性研究综述. *Control and Decision* 控制与决策, 2022, 37: 782–798.

136. Zhu J F 朱军芳, Chen D B 陈端兵, Zhou T 周涛, et al. A survey on mining relatively important nodes in network science 网络科学中相对重要节点挖掘方法综述. *Journal of University of Electronic Science and Technology of China* 电子科技大学学报, 2019, 48: 595–603.

137. Ren X L 任晓龙, Lv L Y 吕琳媛. Review of ranking nodes in complex networks 网络重要节点排序方法综述. *Science Bulletin* 科学通报, 2014, 59(13): 1175–1197.

138. Criado R, Garcí A E, Pedroche F, et al. A new method for comparing rankings through complex networks: Model and analysis of competitiveness of major European soccer leagues. *Chaos: An Interdisciplinary Journal of Nonlinear Science*, 2013, 23(4): 043114

139. Chen W 陈卫. Research on influence diffusion in social network 社交网络影响力传播研究. *Big Data Research* 大数据, 2015, 1(3): 82–98.

140. Liang W, Yan F, Iliyasu A M, et al. Three degrees of influence rule-based Grover walk model with application in identifying significant nodes of complex networks. *Human-Centric Computing and Information Sciences*, 2023, 13: 9.

141. Christakis N A, Fowler J H. Social contagion theory: Examining dynamic social networks and human behavior. *Statistics in Medicine*, 2013, 32(4): 556–577.

142. Xu X K 许小可, Hu H B 胡海波, Zhang L 张伦, et al. Computational Communication on Social Networks 社交网络上的计算传播学. Beijing 北京: Higher Education Press 高等教育出版社, 2015.

143. Chen X L 陈晓龙. Social network influence maximization algorithm and its propagation model research. 社会网络影响力最大化算法及其传播模型研究. Harbin 哈尔滨: Harbin Engineering University 哈尔滨工程大学, 2016.

144. Yang S X 杨书新, Liang W 梁文, Zhu K L 朱凯丽, et al. Measurement of node influence based on three-level neighbor in complex networks 基于三级邻居的复杂网络节点影响力度量方法. *Journal of Electronics & Information Technology* 电子与信息学报, 2020, 42: 1140–1148.

145. Brin S, Page L. The anatomy of a large-scale hypertextual Web search engine. *Computer Networks and ISDN Systems*, 1998, 30(1): 107–117.

146. Han Z M 韩忠明, Chen Y 陈炎, Li M Q 李梦琪, et al. An efficient node influence metric based on triangle in complex networks 一种有效的基于三角结构的复杂网络节点影响力度量模型. *Acta Physica Sinica* 物理学报, 2016, 65: 289–300.

147. Kitsak M, Gallos L K, Havlin S, et al. Identification of influential spreaders in complex networks. *Nature Physics*, 2010, 6(11): 888–893.

148. Rivas A, Huelga S F. *Open quantum systems: An intoduction.* Berlin: Springer, 2012.

149. Tang H, Shi R X, He T S, et al. TensorFlow solver for quantum PageRank in large-scale networks. *Science Bulletin*, 2021, 66(2): 120–126.

150. Rossi L, Torsello A, Hancock E R. Node centrality for continuous-time quantum walks. *Proceedings of the 10th Joint IAPR International Workshops on Statistical Techniques in Pattern Recognition and Structural and Syntactic Pattern Recognition*, 2014. Springer, Joensuu, Finland.

151. Padgett J F, Ansell C K. Robust action and the rise of the medici, 1400-1434. *American Journal of Sociology*, 1993, 98(6): 1259–1319.

152. Bai L, Rossi L, Cui L X, et al. Quantum kernels for unattributed graphs using discrete-time quantum walks. *Pattern Recognition Letters*, 2017, 87: 96–103.

153. Minello G, Rossi L, Torsello A. Can a quantum walk tell which is which? A study of quantum walk-based graph similarity. *Entropy*, 2019, 21(3): 328.

154. Rossi L, Torsello A, Hancock E R. Measuring graph similarity through continuous-time quantum walks and the quantum Jensen-Shannon divergence. *Physical Review E*, 2015, 91(2): 022815.

155. Yan F, Liang W, Hirota K. An information propagation model for social networks based on continuous-time quantum walk. *Neural Computing and Applications*, 2022, 34(16): 13455–13468.

156. Kempe D, Kleinberg J, Tardos E. Maximizing the spread of influence through a social network. *Proceedings of the 9th ACM SIGKDD International Conference on Knowledge Discovery and Data Mining*, 2003. ACM, Washington D.C, USA.

157. Chen W, Wanag Y J, Yang S Y. Efficient influence maximization in social networks. *Proceedings of the 15th ACM SIGKDD International Conference on Knowledge Discovery and Data Mining*, 2009. ACM, Paris, France.

158. Qiu L Q, Gu C M, Zhang S, et al. TSIM: A two-stage selection algorithm for influence maximization in social networks. *IEEE Access*, 2020, 8: 12084–12095.

159. Boito P, Grena R. Quantum hub and authority centrality measures for directed networks based on continuous-time quantum walks. *Journal of Complex Networks*, 2021, 9(6): cnab038.

160. Wu T, Izaac J, Li Z X, et al. Experimental parity-time symmetric quantum walks for centrality ranking on directed graphs. *Physical Review Letters*, 2020, 125(24): 240501.

161. Wang K K, Shi Y H, Xiao L, et al. Experimental realization of continuous-time quantum walks on directed graphs and their application in PageRank. *Optica*, 2020, 7(11): 1524–1530.

162. Cui P, Wang X, Pei J, et al. A survey on network embedding. *IEEE Transactions on Knowledge and Data Engineering*, 2019, 31(5): 833–852.

163. Yu H Y, Braun P, Yildiri M M A, et al. High-quality binary protein interaction map of the yeast interactome network. *Science*, 2008, 322(5898): 104–110.

164. Yang D, Xian J J, Pan L M, et al. Effective edge-based approach for promoting the spreading of information. *IEEE Access*, 2020, 8: 83745–83753.

165. Cheng X Q, Ren F X, Shen H W, et al. Bridgeness: A local index on edge significance in maintaining global connectivity. *Journal of Statistical Mechanics: Theory and Experiment*, 2010, 2010(10): P10011.

166. Li M, Liu R R, Jia C X, et al. Critical effects of overlapping of connectivity and dependence links on percolation of networks. *New Journal of Physics*, 2013, 15(9): 093013.

167. Liang W, Yan F, Iliyasu A M, et al. GCQW: A quantum walk model for predicting missing links of complex networks. *Proceedings of the 2nd IEEE International Conference on Information Communication and Software Engineering* 2022. IEEE, Chongqing, China.

168. Lv L Y 吕琳媛, Zhou T 周涛. *Link Prediction* 链路预测. Beijing 北京: Higher Education Press 高等教育出版社, 2013.

169. Liang W, Yan F, Iliyasu A M, et al. A Hadamard walk model and its application in identification of important edges in complex networks. *Computer Communications*, 2022, 193: 378–387.

170. Holme P, Kim B J, Yoon C N, et al. Attack vulnerability of complex networks. *Physical Review E*, 2002, 65(5): 056109.

171. Liu Y, Tang M, Zhou T, et al. Improving the accuracy of the k-shell method by removing redundant links: From a perspective of spreading dynamics. *Scientific Reports*, 2015, 5(1): 13172.

172. Lü L Y, Chen D B, Ren X L, et al. Vital nodes identification in complex networks. *Physics Reports*, 2016, 650: 1–63.

173. Gupta L, Jain R, Vaszkun G. Survey of important issues in UAV communication networks. *IEEE Communications Surveys & Tutorials*, 2015, 18(2): 1123–1152.

174. Uddin S, Khan A, Piraveenan M. A set of measures to quantify the dynamicity of longitudinal social networks. *Complexity*, 2016, 21(6): 309–320.

175. Newman M E J. A measure of betweenness centrality based on random walks. *Social Networks*, 2005, 27(1): 39–54.

176. Moutinho J P, Melo A, Coutinho B, et al. Quantum link prediction in complex networks. *Physical Review A*, 2023, 107(3): 032605.

177. Zhou T, Lee Y L, Wang G N. Experimental analyses on 2-hop-based and 3-hop-based link prediction algorithms. *Physica A: Statistical Mechanics and its Applications*, 2021, 564: 125532.

178. Pech R, Hao D, Lee Y L, et al. Link prediction via linear optimization. *Physica A: Statistical Mechanics and its Applications*, 2019, 528: 121319.

179. Kovács I A, Luck K, Spirohn K, et al. Network-based prediction of protein interactions. *Nature Communications*, 2019, 10(1): 1240.

180. Liang W, Yan F, Iliyasu A M, et al. A simplified quantum walk model for predicting missing links of complex networks. *Entropy*, 2022, 24(11): 1547.

181. Lorrain F, White H C. Structural equivalence of individuals in social networks. *The Journal of Mathematical Sociology*, 1971, 1(1): 49–80.

182. Salton G, Mcgill M J. *Introduction to Modern Information Retrieval*. Auckland: McGraw-Hill, 1983.

183. Jaccard P. Étude comparative de la distribution florale dans une portion des Alpes et du Jura. *Bulletin de la Société vaudoise des sciences naturelles*, 1901, 37: 547–579.

184. Sorensen T A. A method of establishing groups of equal amplitude in plant sociology based on similarity of species content and its application to analyses of the vegetation on Danish commons. *Biologiske Skrifter*, 1948, 5: 1–34.

185. Ravasz E, Somera A L, Mongru D A. Hierarchical organization of modularity in metabolic networks. *Science*, 2002, 297(5586): 1551–1555.

186. Zhou T, Lü L Y, Zhang Y C. Predicting missing links via local information. *The European Physical Journal B*, 2009, 71(4): 623–630.

187. Xie Y B, Zhou T, Wang B H. Scale-free networks without growth. *Physica A: Statistical Mechanics and its Applications*, 2008, 387(7): 1683–1688.

188. Adamic L A, Adar E. Friends and neighbors on the web. *Social Networks*, 2003, 25(3): 211–230.

189. Katz L. A new status index derived from sociometric analysis. *Psychometrika*, 1953, 18(1): 39–43.

190. Klein D J, Randić M. Resistance distance. *Journal of Mathematical Chemistry*, 1993, 12(1): 81–95.

191. Fouss F, Pirotte A, Renders J- M, et al. Random-walk computation of similarities between nodes of a graph with application to collaborative recommendation. *IEEE Transactions on Knowledge and Data Engineering*, 2007, 19(3): 355–369.

192. Liu X C, Meng D Q, Zhu X Z, et al. Link prediction based on contribution of neighbors. *International Journal of Modern Physics C*, 2020, 31(11): 2050158.

193. Lockhart J, Minello G, Rossi L, et al. Edge centrality via the Holevo quantity. *Proceedings of the 11th Joint IAPR International Workshops on Statistical Techniques in Pattern Recognition and Structural and Syntactic Pattern Recognition*, 2016. Springer, Mérida, Mexico.

194. Fortunato S. Community detection in graphs. *Physics Reports*, 2010, 486(3–5): 75–174.

195. Li X J 李晓佳, Zhang P 张鹏, Di Z R 狄增如, et al. Community structure in complex networks 复杂网络中的社团结构. *Complex Systems and Complexity Science* 复杂系统与复杂性科学, 2008, 18: 19–42.

196. Lancichinetti A, Fortunato S. Limits of modularity maximization in community detection. *Physical Review E*, 2011, 84(6): 066122.

197. Whang J J, Gleich D F, Dhillon I S. Overlapping community detection using seed set expansion. *Proceedings of the 22nd ACM International Conference on Information & Knowledge Management*, 2013. ACM, New York, USA.

198. Li Y Y 李阳阳, Jiao L C 焦李成, Zhang D 张丹, et al. *Quantum Computing Intelligence* 量子计算智能. Xian 西安: Xidian University Press 西安电子科技大学出版社, 2019.

199. Li L L, Jiao L C, Zhao J Q, et al. Quantum-behaved discrete multi-objective particle swarm optimization for complex network clustering. *Pattern Recognition*, 2017, 63: 1–14.

200. Pizzuti C. GA-Net: A genetic algorithm for community detection in social networks. *Proceedings of the 10th International Conference on Parallel Problem Solving from Nature*, 2008. Springer, Dortmund, Germany.

201. Rozemberczki B, Davies R, Sarkar R, et al. GEMSEC: Graph embedding with self clustering. *Proceedings of the 2019 IEEE/ACM International Conference on Advances in Social Networks Analysis and Mining*, 2019. ACM, Vancouver, Canada.

202. Pons P, Latapy M. Computing communities in large networks using random walks. *Proceedings of the 20th International Symposium on Computer and Information Sciences*, 2005. Springer, Istanbul, Turkey.

203. Blondel V D, Guillaume J- L, Lambiotte R, et al. Fast unfolding of communities in large networks. *Journal of Statistical Mechanics: Theory and Experiment*, 2008, 2008(10): P10008.

204. Clauset A, Newman M E J, Moore C. Finding community structure in very large networks. *Physical Review E*, 2004, 70(6): 066111.

205. Traag V A, Waltman L, Van Eck N J. From Louvain to Leiden: Guaranteeing well-connected communities. *Scientific Reports*, 2019, 9(1): 5233.

206. Traag V A, Krings G, Van Dooren P. Significant scales in community structure. *Scientific Reports*, 2013, 3(1): 2930.

207. Kozdoba M, Mannor S. Community detection via measure space embedding. *Proceedings of the 29th International Conference on Neural Information Processing Systems*, 2015. ACM, Montreal, Canada.

208. Reichardt J, Bornholdt S. Statistical mechanics of community detection. *Physical Review E*, 2006, 74(1): 016110.

209. Paré S F, Gasulla D G, Vilalta A, et al. Fluid communities: A competitive, scalable and diverse community detection algorithm. *Proceedings of the 7th International Conference on Complex Networks and their Applications*, 2018. Springer, Cambridge, UK.

210. Biemann C. Chinese whispers-an efficient graph clustering algorithm and its application to natural language processing problems. *Proceedings of TextGraphs: The 1st Workshop on Graph Based Methods for Natural Language Processing*, 2006. ACL, New York, USA.

211. Newman M E J, Leicht E A. Mixture models and exploratory analysis in networks. *Proceedings of the National Academy of Sciences*, 2007, 104(23): 9564–9569.

212. Higham D J, Kalna G, Kibble M. Spectral clustering and its use in bioinformatics. *Journal of Computational and Applied Mathematics*, 2007, 204(1): 25–37.

213. Rosvall M, Bergstrom C T. Multilevel compression of random walks on networks reveals hierarchical organization in large integrated systems. *PLoS One*, 2011, 6(4): e18209.

214. Brassard G. Searching a quantum phone book. *Science*, 1997, 275(5300): 627–628.

215. Ma H, Yang H X, Lyu M R, et al. Mining social networks using heat diffusion processes for marketing candidates selection. *Proceedings of the 17th ACM Conference on Information and Knowledge Management*, 2008. ACM, Napa Valley, USA.

216. Zhang J W, Philip S Y. *Broad learning Through Fusions*. Berlin: Springer, 2019.

217. Yang S X 杨书新, Liang W 梁文, Zhu K L 朱凯丽. Reverse influence maximization algorithm in social networks 社交网络中对立影响最大化算法. *Journal of Computer Applications* 计算机应用, 2020, 40: 1944–1949.

218. Liu Z Y, Zhou J. *Introduction to Graph Neural Networks*. Berlin: Morgan & Claypool Publishers, 2020.

219. Schuld M, Sinayskiy I, Petruccione F. Quantum walks on graphs representing the firing patterns of a quantum neural network. *Physical Review A*, 2014, 89(3): 032333.

220. Zhang Z H, Chen D D, Wang J J, et al. Quantum-based subgraph convolutional neural networks. *Pattern Recognition*, 2019, 88: 38–49.

221. Rossi R A, Ahmed N K. Role discovery in networks. *IEEE Transactions on Knowledge and Data Engineering*, 2014, 27(4): 1112–1131.

222. Ren P, Aleksić T, Emms D, et al. Quantum walks, Ihara zeta functions and cospectrality in regular graphs. *Quantum Information Processing*, 2011, 10(3): 405–417.

223. Bai L, Ren P, Hancock E R. A hypergraph kernel from isomorphism tests. *Proceedings of the 22nd International Conference on Pattern Recognition*, 2014. IEEE, Stockholm, Sweden.

224. Ahmed N K, Rossi R, Lee J B, et al. Learning role-based graph embeddings. arXiv preprint arXiv:180202896, 2018.

225. Grover A, Leskovec J. Node2vec: Scalable feature learning for networks. *Proceedings of the 22nd ACM SIGKDD International Conference on Knowledge Discovery and Data Mining*, 2016. ACM, San Francisco, USA.

226. Henderson K, Gallagher B, Li L, et al. It's who you know: Graph mining using recursive structural features. *Proceedings of the 17th ACM SIGKDD International Conference on Knowledge Discovery and Data Mining*, 2011. ACM, San Diego, USA.

227. Henderson K W, Gallagher B, Eliassi-Rad T, et al. RolX: Structural role extraction & mining in large graphs. *Proceedings of the 18th ACM SIGKDD International Conference on Knowledge Discovery and Data Mining*, 2012. ACM, Beijing, China.

228. Donnat C, Zitnik M, Hallac D, et al. *Learning Structural Node Embeddings via Diffusion Wavelets*. New York: ACM, 2018.

229. Dernbach S, Mohseni-Kabir A, Pal S, et al. Quantum walk neural networks with feature dependent coins. *Applied Network Science*, 2019, 4(1): 76.

230. Kipf T N, Welling M. Semi-supervised classification with graph convolutional networks. arXiv preprint arXiv:160902907, 2016.

231. Atwood J, Towsley D. Diffusion-convolutional neural networks. *Proceedings of the 30th Conference on Neural Information Processing Systems*, 2016. CAI, Barcelona, Spain.

232. Veličković P, Cucurull G, Casanova A, et al. Graph attention networks. arXiv preprint arXiv:171010903, 2017.

233. Feragen A, Kasenburg N, Petersen J, et al. Scalable kernels for graphs with continuous attributes. *Proceedings of the 26th Conference on Neural Information Processing Systems*, 2013. CAI, Nevada, USA.

234. Borgwardt K M, Kriegel H P. Shortest-path kernels on graphs. *Proceedings of the 5th IEEE International Conference on Data Mining*, 2005. IEEE, Houston, USA.

235. Shervashidze N, Vishwanathan S, Petri T, et al. Efficient graphlet kernels for large graph comparison. *Proceedings of the 12th International Conference on Artificial Intelligence and Statistics*, 2009. PMLR, Florida, USA.

236. Niepert M, Ahmed M, Kutzkov K. Learning convolutional neural networks for graphs. *Proceedings of the 33rd International Conference on Machine Learning*, 2016. PMLR, New York, USA.

237. Zhang M H, Cui Z C, Neumann M, et al. An end-to-end deep learning architecture for graph classification. *Proceedings of the 32nd AAAI Conference on Artificial Intelligence*, 2018. AAAI, New Orleans, USA.

238. Bai L, Rossi L, Zhang Z H, et al. An aligned subtree kernel for weighted graphs. *Proceedings of the 32nd International Conference on Machine Learning*, 2015. PMLR, Lille, France.

239. Bai L, Zhang Z H, Ren P, et al. An edge-based matching kernel through discrete-time quantum walks. *Proceedings of the 7th International Conference on Image Analysis and Processing*, 2015. Springer, Genoa, Italy.

Index

Note: **Bold** page numbers refer to tables, *italic* page numbers refer to figures.